Regiane Godoy de Lima

Superconductor through microwave-assisted hydrothermalization

Regiane Godoy de Lima

Superconductor through microwave-assisted hydrothermalization

Bismuth-based ceramic superconductor produced by the Pechini Method with microwave-assisted hydrothermalization

ScienciaScripts

Imprint

Cover image: www.ingimage.com

This book is a translation from the original published under ISBN 978-3-330-75638-0.

Publisher:
Sciencia Scripts
is a trademark of
Dodo Books Indian Ocean Ltd. and OmniScriptum S.R.L publishing group

120 High Road, East Finchley, London, N2 9ED, United Kingdom
Str. Armeneasca 28/1, office 1, Chisinau MD-2012, Republic of Moldova, Europe
Printed at: see last page
ISBN: 978-620-8-27764-2

SUMMARY

To my mother, for everything

ACKNOWLEDGMENTS

To my family, for my upbringing, especially my mother Marta Regina Godoy, who taught me values that I will carry with me for life, and whose main goal in life is to see her children happy and fulfilled. And she always supported me in everything I wanted. To my father Herminio de Lima, for understanding me, supporting me and always looking out for me wherever I was. To my siblings: Mariane Godoy de Lima, for always supporting me and giving me strength, and Vitor de Lima for living with me, for his patience, wisdom and for playing basketball with me.

To Prof. Rafael Zadorosny, for guiding me, welcoming me, believing in my professional capacity, giving me his time and valuable advice.

To Prof. Clàudio Luiz Carvalho, for his attention to the work, always with discussions to improve the work and pertinent questions.

To Prof. Angel Fidel Vilche for his friendship and for starting me in the academic and research fields, aiming for my growth, and giving me support and advice for life.

To Prof. Silvio Rainho Teixeira for his support in all my work since graduation, always with valuable contributions and help in synthesizing and using the micro-wave assisted hydrothermal method.

To Prof. Agda Eunice de Souza, for having accepted the invitation to be part of the examining committee for the work, and for her contribution to the research, help in the synthesis and for the discussions of the work in order to obtain a better result.

To Prof. Fabiano Colauto for accepting the invitation to take part in the examining committee.

To the professors and staff of the DFQ for their academic support.

To my friends at GDAM, Guilherme Bortega, Gustavo Quereza Freitas, Joao Borges Silveira, for the discussions, work growth and coffee. Especially Gisele Aparecida de Souza for her help with measurements, teachings and discussions. Elwis Carlos Sartorelli Duarte, for all the discussions on superconductivity and other subjects. Vivian Rodrigues Delmute, for instructing me from the start of the work, with discussions and opinions that were essential for my professional growth, and discussions on life issues.

To my undergraduate and postgraduate colleagues, for their companionship and help whenever necessary.

To Beatriz Barra, for her companionship and friendship, who has always been by my side since graduation, sharing the same feelings.

To lifelong friends Patricia Felipe, Mônica Moraes and Carolina Preto de Godoy for always being by my side, no matter how far away I am.

To Michel Jones da Silva, Ricardo Hidalgo Santin and Alex Otàvio Sanches for the XRD measurements.

To Élton José de Souza, for his SEM and EDS measurements.

To the Superconductivity and Magnetism Group of the Federal University of Sao Carlos for the magnetic measurements.

And finally, to CAPES and CNPq for their financial support.

"We look at mountains and say that they are eternal, and that's what they seem to be... But in the course of time, mountains rise and fall, rivers change course, stars fall from the sky, and great cities sink into the sea. We think that even the gods die. Everything changes."

(George R.R. Martin - Fury of Kings)

SUMMARY

The synthesis routes for superconducting cuprates seek to improve their properties, such as obtaining single phases, homogeneous and better connected grains which, consequently, interfere with increasing the capacity to carry electric current. In this work we studied the synthesis of the superconducting oxide system Bi:Sr:Ca:Cu:O in phase 2212 (BSCCO-2212), which has a critical temperature of around 85 K. To synthesize this material, a variation of the polymer precursor method developed by Pechini was used and, in a second route, a step was added to this method in which the microwave-assisted hydrothermalization method was applied. In the latter, the material is heated from the inside out and, due to the interaction of the microwaves with the solution, the temperature of the system increases rapidly, directly influencing the speed of synthesis. In contrast, in conventional heating, the material is first heated on the surface and only then is heat transferred to the inside of the system. The powders resulting from the two synthesis methods described above were heat-treated and then pressed to obtain pellets. Structural, electrical and magnetic characterizations were then carried out in order to study the influence of each method on the properties of the synthesized materials.

Keywords: Superconductor. Pechini. Hydrothermalization. Microwaves.

1 INTRODUCTION

The use of microwaves in processing and obtaining materials has grown in recent years in various fields of knowledge, such as chemistry, physics and materials engineering. In some studies, the use of domestic microwave ovens in scientific activities has aroused interest due to a series of new applications, simplicity and low cost (KEYSON et al., 2006). For this work, the microwave-assisted hydrothermalization method coupled with the Pechini method was used to prepare a superconducting ceramic material. There are currently several methods for preparing superconducting ceramic materials, such as Bi2Sr2CaCu2Ox (BSCCO-2212). Chemical methods such as sol-gel, Pechini, precipitation and hydrothermalization methods are used to overcome the limitations of the conventional method, i.e. the solid-state reaction, as they allow for better mixing of the constituent cations, forming a homogeneous material (RAO; NAGARAJAN; VIJAYARAGHAVEN, 1994; ZHANG et al., 2010). The microwave-assisted hydrothermalization method was used because the heating takes place inside the copper oxides and from there is transferred to the surroundings, allowing the crystallization temperature to be reached quickly, and consequently saving on electricity costs and synthesis time (VOLANTI et al., 2007). In heating by the conventional hydrothermalization method, unlike microwave-assisted hydrothermalization, the synthesized material begins to heat up from the surface and this transfers thermal energy to the oxides being processed (VOLANTI et al., 2007). It should be noted that, to our knowledge, there is no work in the literature using the microwave-assisted hydrothermalization method to synthesize superconducting oxides.

The general objective of this work was to study the influence of the Pechini and microwave-assisted hydrothermalization methods on the synthesis of the superconducting ceramic material BSCCO in the stoichiometric form Bi2Sr2CaCu2Ox, thus evaluating the structural, electrical and magnetic properties of the materials produced.

Specifically, samples were produced by the Pechini method in association with the microwave-assisted hydrothermal method, using different metal salts to prepare the precursor solution. The different samples were then analyzed and compared in terms of their electrical, magnetic and structural properties.

This dissertation is divided into five sections, in addition to this introduction, each of which is divided as follows:

2. Literature Review: A brief introduction to superconductivity and some properties of

superconducting materials is presented. Some properties and characteristics of the BSCCO system are also discussed, as well as a description of phase diagrams and synthesis methods.

3. Materials and Methods: Describes the procedures used to synthesize the samples, such as the chemical routes, the hydrothermal method and the heat treatments. This section also presents a summary of the experimental methods used to characterize the samples.

4. Results and Discussions: This section contrasts and discusses the results of the synthesized samples. The data obtained for the different samples will be presented, as well as the characterizations carried out, such as X-ray diffraction (XRD), scanning electron microscopy (SEM), energy dispersive X-ray spectroscopy (EDS), electrical and magnetic spectroscopy.

5. Conclusion: Presentation of the observations and, consequently, the conclusions drawn from the studies carried out.

2 LITERATURE REVIEW

2.1 SUPERCONDUCTIVITY

The phenomenon of superconductivity was first observed in 1911 in Leiden, Holland, by H. Kamerling Onnes. Three years earlier, he had liquefied helium, reaching a temperature of 4.2 K. When Onnes was studying a sample of mercury, he observed that the resistivity of his sample dropped abruptly to zero around 4.2 K. This temperature became known as the critical temperature, T_c . Figure 1 shows the curve obtained by Onnes. With this discovery, a new class of materials emerged, the so-called superconductors (BUCKEL; KLEINER, 2004; KITTEL, 2005).

Figure 1. Resistance of a mercury sample as a function of temperature.

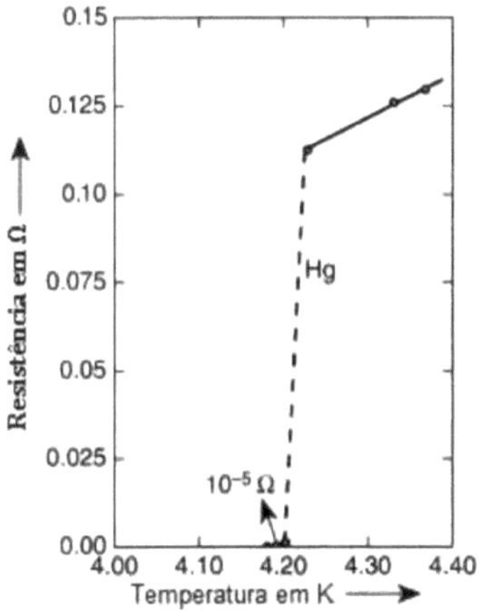

Source: Adapted from Buckel and Kleiner (2004).

2.2 SUPERCONDUCTING STATE

Superconductivity is characterized by three distinct properties. The best known is the zero electrical resistance at temperatures below the critical value T_c, typical of each material, i.e. it is a state where the transport of electric current is done without losses due to the Joule Effect (BUCKEL; KLEINER, 2004; POOLE JR. et al., 2007).

The second characteristic is the discontinuity of the specific heat, as shown in Figure 2. The transition between the normal and superconducting states, in the absence of an applied magnetic field, is a second-order phase transition. This means that there is no latent heat of transition (POOLE JR. et al., 2007). Above T_c, the specific heat of these materials tends to follow Debye's theory (C_n =

γT+AT³), where the linear term in T occurs due to the contribution of conduction electrons and T^3 is due to the contribution of vibrations in the crystal lattice. The constants *y* and *A* are characteristic of each material.

Figure 2. Discontinuity in the specific heat of an Al sample when it reaches T_c compared to the specific heat in the normal state.

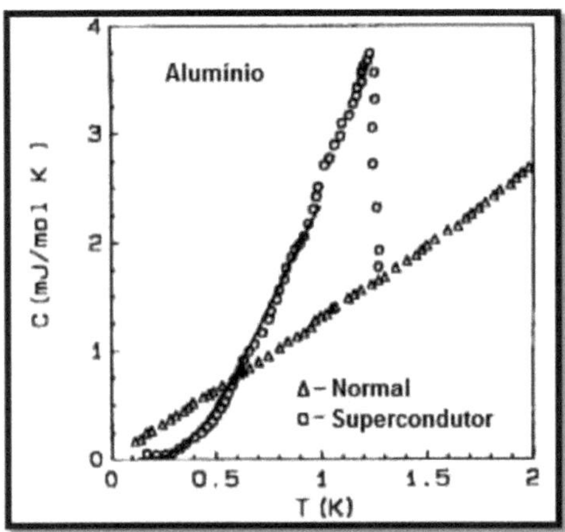

Source: Adapted from Poole Jr. et al. (2007).

Perfect diamagnetism is another characteristic of superconductivity. This was discovered in 1933 by the German physicists Meissner and Ochsenfeld. They noticed that when a superconductor is subjected to a magnetic field and then cooled below T_c, the magnetic flux is expelled from its interior. This property became known as the Meissner Effect and is shown in Figure 3 (CALLISTER JUNIOR, 2008; KITTEL, 2005).

Figure 3 - Schematic of the Meissner effect. The magnetic field lines are excluded from a superconductor when it is below the critical temperature.

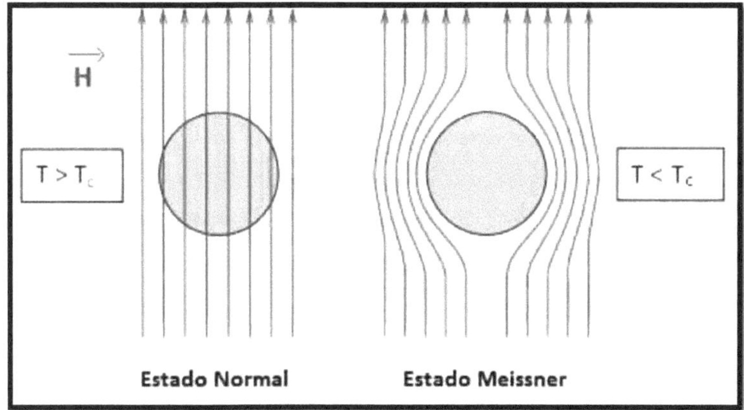

Source: Adapted from Callister Junior (2008).

Since the discovery of the Meissner Effect, theories have emerged in an attempt to explain the effect of a magnetic field on the phenomenon of superconductivity. We will briefly describe three of them in this paper, two of which are phenomenological constructions, the London brothers' theory and the Ginzburg-Landau theory, and a third, based on first principles, known as the BCS theory.

2.3 theories ON SUPERCONDUCTIVITY

2.3.1 London equation

The London brothers' theory is based on the electrodynamic properties of superconductors and anticipated the existence of a magnetic field penetration length, λ_L, which depends on each material (BUCKEL; KLEINER, 2004; KITTEL, 2005) and is given by Equation 1:

$$\lambda_L = \left(\frac{m}{\mu_0 n_s e^2}\right)^{1/2} \tag{1}$$

where m is the effective mass of the electron, μ_0 is the magnetic permeability of the vacuum, n_s is the density of superconducting electrons and *e* is the charge of the electron.

The London equations provided a semi-quantitative description of electrodynamic phenomena. Except for the accumulation of experimental data, no major steps were taken before the emergence of the Ginzburg-Landau theory (CYROT, 1973).

2.2.1 Ginzburg-Landau theory

In 1950, Soviet physicists V. L. Ginzburg and L. D. Landau formulated a phenomenological theory to explain the thermodynamic properties of the transition from the normal state to the superconducting state (BALSEIRO; CRUZ, 1988) in the presence of an applied magnetic field. Ginzburg and Landau pointed out that London's phenomenological theory was unsatisfactory from two points of view. Firstly, it was not possible to determine the surface tension at the boundary between the normal and superconducting states. Secondly, the surface energy linked to the field and the supercurrent, as predicted from the solution of the London equation, was negative, unlike the observed surface energy, which was positive. Therefore, in order to obtain this positive energy, it was necessary to introduce a surface energy of non-magnetic origin, which is greater than that linked to the distribution of the field. The theory also failed to describe the destruction of superconductivity by a current because it considered the number of superconducting electrons, n_s, only as a function of temperature. Ginzburg and Landau's aim was to build a more comprehensive theory. In doing so, they introduced new features into the formulation of their theory, which allowed for important new developments in the decade that followed (CYROT, 1973).

Ginzburg and Landau introduced an order parameter ψ to describe the phenomena of superconductivity. In the superconducting phase, which is the ordered phase, $\psi \neq 0$, while for temperatures above T_c, $\psi = 0$ when thermodynamic equilibrium is reached. Ginzburg and Landau started from the idea that ψ was an effective wave function of the superconducting electrons, such that their density, present in the London equations, was given by the local relation $n_s = |\psi(x)|^2$. They then assumed an expansion of the free energy in powers of ψ.

The central problem of Gizburg-Laundau smoothing is to find the functions ψ(x,y,z) and $\vec{A}$ (x,y,z) ($\vec{A}$ is the potential vector) that minimize the total free energy of the system. Thus, by minimizing the energy with respect to ψ and $\vec{A}$ we obtain the two Ginzburg-Landau equations. From these equations we can extract important parameters which will be discussed in section 2.4.

It is worth noting that, at the time of its publication, this theory, being simple and phenomenological, was little appreciated (CYROT, 1973) by researchers in the field. However, its formulation was not the one we use today.

2.2.2 BCS theory

In 1957, physicists J. Bardeem, L.N. Copper and T.R. Schrieffer formulated a microscopic theory for superconductivity, known as BCS. This theory is based on the fact that superconductivity occurs due to the strong interaction of electrons with the lattice of the material, i.e. an electron moving through a lattice interacts with its neighboring positive ions, which oscillate about their equilibrium positions. Thus, the charge distortion resulting from this interaction propagates along the lattice, causing distortions in the periodic potential. This distortion can influence the movement of another electron that is at a certain distance from this disturbed region and which also interacts with the oscillating lattice. This propagation of lattice vibrations is quantized and is called a phonon. This interaction, i.e. *electron-phonon*, is shown schematically in Figure 4. Two electrons interacting with each other via the intermediate phonon can form bound states and this pair of bound electrons is called a Cooper pair. These are the charge carriers in superconductors, also called superelectrons, which in turn, when they move, establish the so-called supercurrents (BUCKEL; KLEINER, 2004; OSTERMANN; PUREUR, 2005; POOLE JR. et al., 2007). The average size of this electronic bond is characterized by the coherence length (ξ).

Figure 4 - Illustration of the passage of an electron, with its negative charge, through the crystal lattice formed by positive ions, which causes them to move away from their equilibrium position. This effect is linked to the emergence of attractive forces between pairs of electrons, the Cooper pairs.

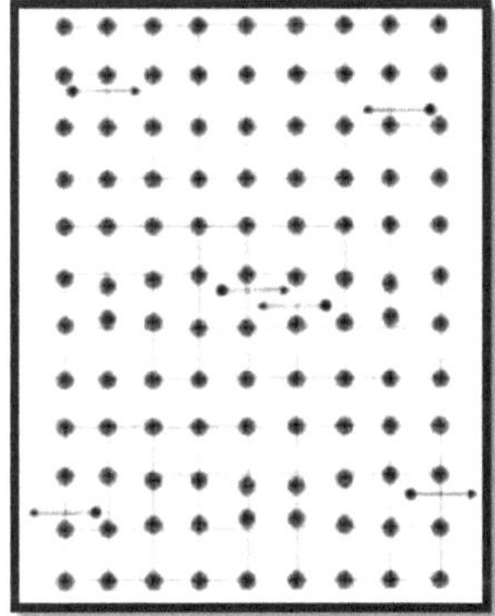

Source: Ostermann and Pureur (2005).

2.4 TYPE I, TYPE II AND MIXED STATE SUPERCONDUCTORS

In 1957, Abrikosov published a study in which, working with the Ginzburg-Landau theory, he demonstrated that it was possible, under certain conditions, for magnetic flux to penetrate the interior of some superconductors. Naturally, from the Ginzburg-Landau theory

Landau extracts the two characteristic lengths of superconductors, i.e. λ and ξ. The ratio between these two lengths defines the Ginzburg-Landau parameter $\kappa = \frac{\lambda}{\xi}$ (KITTEL, 2005). Depending on the value of κ, the surface energy density can acquire positive or negative values and, consequently, characterizes the two types of superconductors, type I (SC I) with values of $\kappa < (1/\sqrt{2})$ for which the surface energy is positive, and type II (SC II), with values of *κ >(1/√2)*, for which the surface energy is negative (KITTEL, 2005; POOLE JR. et al., 2007).

The two types of superconductor have similar thermal properties when transitioning to the normal state in the absence of a magnetic field. However, their behavior becomes quite different when a magnetic field is applied to the material (KITTEL, 2005).

SCIs completely exclude an applied magnetic field from their interior until it reaches a critical value above which superconductivity is destroyed, i.e. from this point onwards, the field penetrates the sample completely. In the case of an SCII, the complete Meissner effect occurs up to a field value called the lower critical field, H_{c1} . Above H_{c1} , the magnetic flux penetrates the material and a mixed state is formed, where the normal state, with penetrated flux, and the superconducting state coexist. The superconductivity will be destroyed when the external field reaches a value known as the upper critical field, H_{c2} . The magnetic response described above is exemplified in Figure 5. In some cases there may still be superconductivity remaining on the surface of the sample and this will be destroyed when the applied field reaches the value known as H_{c3}, which is usually $1.7H_{c2}$ (KITTEL, 2005).

Figure 5: Magnetization as a function of the applied magnetic field for (a) a SCI and (b) SCII.

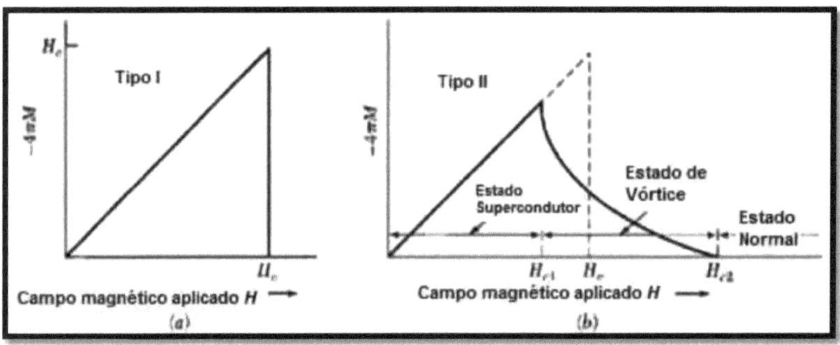

Source: Adapted from Kittel (2005).

The magnetic flux that penetrates the SCIIs does so in a singular way, i.e. the penetration is quantized. Each element is called a **vertex**, which is why the mixed state is also known as a **vertex state**.

The vortices can be thought of as very long tubes, oriented parallel to the direction of the induction lines of the applied field. In this region, superconductivity is eliminated. Shielding supercurrents circulate around these tubes, preventing the magnetic field contained in their core from spreading throughout the superconducting sea. As the interaction between the vortices is repulsive, when in high density, they arrange themselves to form a hexagonal network, called an Abrikosov network, as illustrated in Figure 6 (BUCKEL; KLEINER, 2004; HOFFMAN, 2012; POOLE JR. et al., 2007).

Figure 6 - Magnetic field flow surrounded by the superconduction currents forming the hexagonal Abrikosov network.

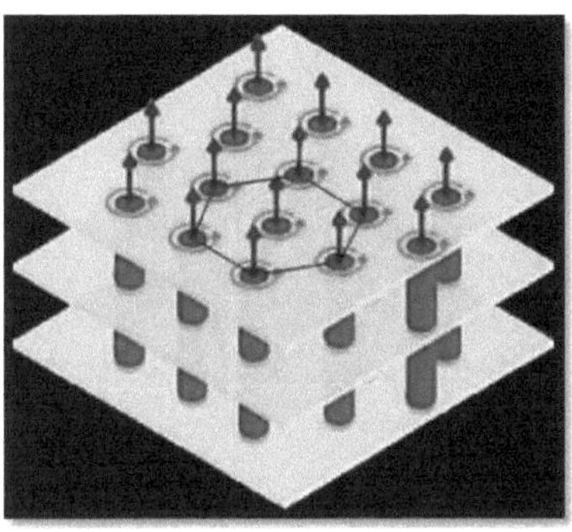

Source: Hoffman (2012).

2.5 ELECTRICAL PROPERTIES

When a potential difference exists between two points on a wire of conductive material, a uniform electric field *E* is established along the axis of the wire. This field exerts a force that accelerates the electrons in the material (OSTERMANN; PUREUR, 2005; POOLE JR. et al., 2007), that is,

$$F = -eE \tag{2}$$

But,

$$F = m.a \tag{3}$$

So,

$$-eE = m\left(\frac{dv}{dt}\right) \tag{4}$$

$$v = -\left(\frac{eE}{m}\right)\tau \tag{5}$$

where τ is the average time between collisions of an electron, v is the drift velocity, and e and m are, respectively, the charge and mass of the electron.

The movement of electrons consists of successive periods of acceleration interrupted by collisions and, on average, each collision reduces the velocity of electrons to zero before the next acceleration begins (POOLE JR. et al., 2007).

The current density can be written in terms of the drift velocity.

$$\vec{J} = ne\vec{v} \tag{6}$$

where $\vec{v}$ is given by equation (5). This gives us

$$\vec{J} = \left(\frac{ne^2\tau}{m}\right)\vec{E} \tag{7}$$

The constant in brackets in equation (7) is known as the electrical conductivity and so we can write:

$$\vec{J} = \sigma\vec{E} \tag{8}$$

With

$$\sigma = \frac{ne^2\tau}{m} \tag{9}$$

is the electrical conductivity. We can rewrite equation (8) in terms of the material's resistivity.

$$\vec{E} = \rho\vec{J} \tag{10}$$

Where p=1/σ is called electrical resistivity.

Equations (8) and (10) represent Ohm's law, which basically refers to the linear dependence between electric field and electronic current. Many metals obey this law.

As we can see from equation (9), the conductivity depends on *ne,* which is the charge density available for transport; it also depends on the *e/mc* ratio, i.e. the acceleration produced by the electric field, and it also depends directly on the average collision time of the electrons, τ. It is during this relaxation time that the electron acquires energy from the field. Therefore, the longer τ, the higher the metal's conductivity and the lower its electrical resistivity (OSTERMANN; PUREUR, 2005).

Electron collisions with lattice vibrations give rise to a temperature-dependent contribution to the relaxation time (OSTERMANN; PUREUR, 2005; POOLE JR. et al., 2007). Since resistivity is inversely proportional to τ, its behavior as a function of temperature in a metallic system can be made up of two contributions, namely,

$$\rho(T) = \rho_0 + \rho_i(T) \tag{11}$$

the term p_0 is called residual resistivity and is due to electronic collisions with static imperfections in the lattice. This term is independent of temperature and only becomes dominant at temperatures close to absolute zero. The $p_i(T)$ term is the ideal resistivity, which is mainly due to electronic interactions with vibrations in the crystal lattice (phonons). Figure 7 (OSTERMANN; PUREUR, 2005; POOLE JR. et al., 2007) shows a curve that exemplifies what has been described.

Figure 7. Electrical resistance of a metal as a function of temperature.

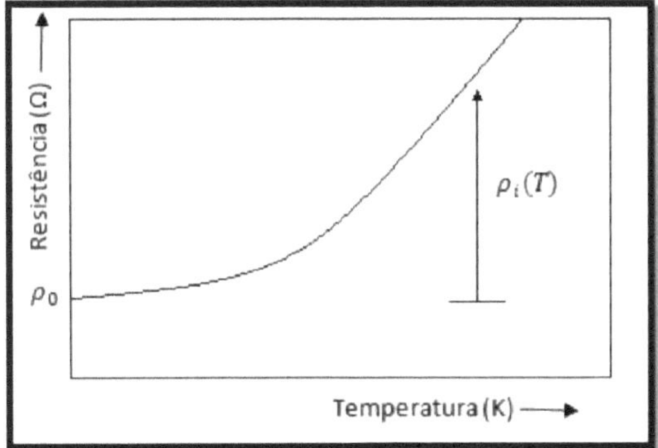

Source: Adapted from Ostermann and Pureur (2005).

The term p_i *(T),* at high temperature, is linear with *T* and, at low temperatures, i.e. well below the Debye temperature (Θ_D), according to Bloch's Law, it is proportional to T^5, i.e. $\sigma \propto T^{-5}$ (POOLE JR. et al., 2007).

Classical or low-temperature superconductors follow Bloch's law when in the normal state, i.e. dependence on T^5 if the transition temperature is very low, as illustrated in Figure 8(a). High-temperature superconductors have T_c's in the linear region, corresponding to the resistivity shown in Figure 8(b), (POOLE JR. et al., 2007).

Figure 8. Abrupt drop in resistivity to zero at the transition temperature (T_c) (a) for a low-temperature superconductor in the Bloch region of T^5 and (b) for a high-temperature superconductor in the linear region.

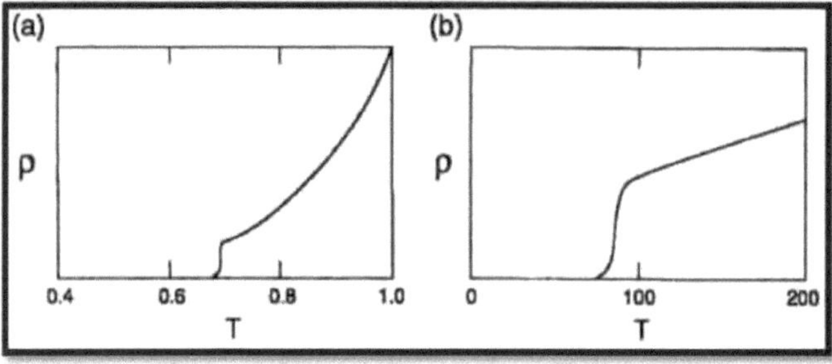

Source: Poole Jr. et al. (2007).

2.6 MAGNETIC PROPERTIES

As described above, superconducting materials are perfect diamagnets, which means that their magnetic susceptibility is negative, i.e. χ = -1, in the international system of units.

The demonstration is simple, defining $\chi = \frac{\vec{M}}{\vec{H}}$ and taking the constitutive relationship between the fields:

$$\vec{B} = \mu_0(\vec{H} + \vec{M}), \tag{12}$$

we have

$$\vec{B} = \mu_0\vec{H}(1 + \chi), \tag{13}$$

where $\vec{H}$ *e* $\vec{M}$ are respectively the applied field and the magnetization and μ_0 is the magnetic permittivity of the vacuum.

In the Meissner state, the field inside the material is zero, i.e, $\vec{B}$ = 0, so from (12) and (13) we have that:

$$\vec{M} = -\vec{H} \tag{14}$$

and therefore,

$$\chi = -1 \tag{15}$$

Again, we call this the Meissner effect. Thus, when a magnetic field is applied to a superconductor, it is shielded from the external field (POOLE JR. et al., 2007).

In order to differentiate between a superconductor and a perfect conductor, we will analyze two processes of subjecting a material to an external magnetic field. The first process, called *Zero Field Cooled* (ZFC), leads to the exclusion of flux, i.e. if a material in its normal state is cooled below T_c and only then is an external magnetic field applied, it will be excluded from the superconductor. The second procedure, called *Field Cooled* (FC), refers to the expulsion of the flux, i.e. a material in its normal state is placed in a magnetic field, with the field penetrating homogeneously, because the permeability μ of the material is close to the permeability of vacuum, μ_0 . In the presence of this field, the material cools down below T_c , and the field is then expelled from the material. It is this process that characterizes the Meissner effect, i.e. what differentiates a superconductor from a perfect conductor. Figure 9 (POOLE JR. et al., 2007) shows a comparison of the responses of a homogeneous superconductor, a superconductor with a central hole and a perfect conductor when subjected to the ZFC and FC procedures.

Figure 9. Zero Faithful Cooled (ZFC) and Field Cooled (FC) procedures on a solid Superconducting cylinder (left), a Superconducting cylinder with an axial hole (center), and a perfect conductor (right).

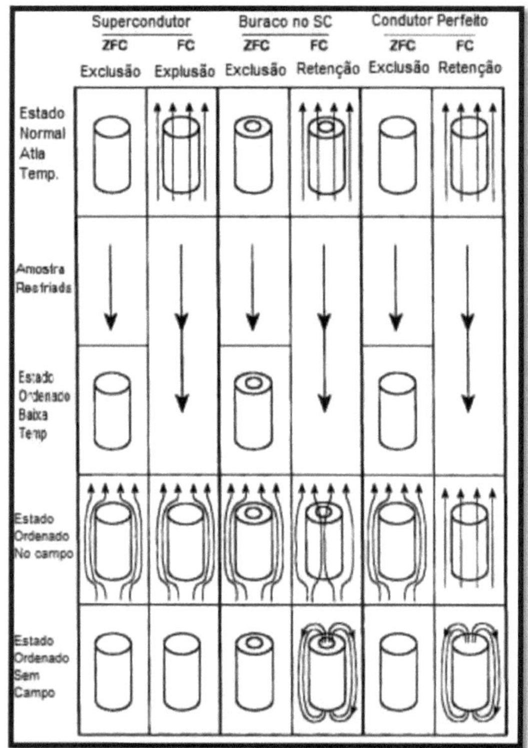

Fonte: Adaptado de Poole Jr. et al. (2007).

2.7 CHRONOLOGICAL DEVELOPMENT OF T_c SUPERCONDUCTING MATERIALS

In 1973, the American physicist B. Matthias discovered that the compound Nb_3 Ge, was superconducting with T_c = 23 K. This was, for many years, the critical temperature record for a superconducting material. In January 1986, in Switzerland, K. A. Muller and J.G. Bednorz demonstrated that the material $La_{2-x}Ba_xCuO$ (LBCO) became superconducting at 30 K (BALSEIRO; CRUZ, 1988; BEDNORZ; MÜLLER, 1986; BUCKEL; KLEINER, 2004; POOLE JR. et al., 2007). In 1987 they were awarded the Nobel Prize for Physics. This discovery instigated the search for materials with higher critical temperatures. In the same year, physicists Paul Chu and Maw-Kuen Wu discovered a system composed of Y-Ba-Cu-O (YBCO) with a critical temperature of 92 K (BALSEIRO; CRUZ, 1988). In addition to an increase in T_c (compared to LBCO), this temperature is higher than the boiling temperature of liquid nitrogen (77 K), which represented a major advance in terms of material cooling costs, opening up new horizons for technological applications (BALSEIRO; CRUZ, 1988).

In early 1988, two new superconducting systems based on bismuth and thallium were discovered, with transition temperatures considerably higher than those reached with YBaCuO. These compounds have approximately the same lattice parameters *a* and *b* as Y and La, but with much larger *c* dimensions. These compounds are $Bi_2 Sr_2 Ca_n Cu_{n+1} O_{6+2n}$ and $Tl_2 Ba_2 Ca_n Cu_{n+1} O_{6+2n}$, with n=0, 1, 2 and 3, which is related to the number of planes formed by the copper oxide. These planes are responsible for supplying the material with charge carriers (POOLE JR. et al., 2007).

The Tl system was discovered by Sheng and Hermam in 1988, which also had T_c in the 25K to 122K range (Tl-1223). The $HgBa_2 Ca_{n-1}Cu O_{2n+2+\delta}$ system (n=3, T_c = 133K) was discovered by Schilling in the 1990s (POOLE JR. et al., 2007).

In March 2001, a new superconducting compound was discovered, MgB_2, with a Tc of around 39 K (MARTINEZ, 2005). This new superconductor differs from copper oxide-based superconductors in that it is an intermetallic compound.

In 2008, superconductivity was discovered in materials based on iron oxide with rare earth elements, the so-called pinictides. These have very interesting superconducting characteristics (KAMIHARA et al., 2008). Figure 10 shows the chronological evolution of Tc since the discovery of superconductivity in 1911 (DIAS, 2011; POOLE JR. et al., 2007).

Figure 10. Chronological evolution of the Tc of superconducting materials.

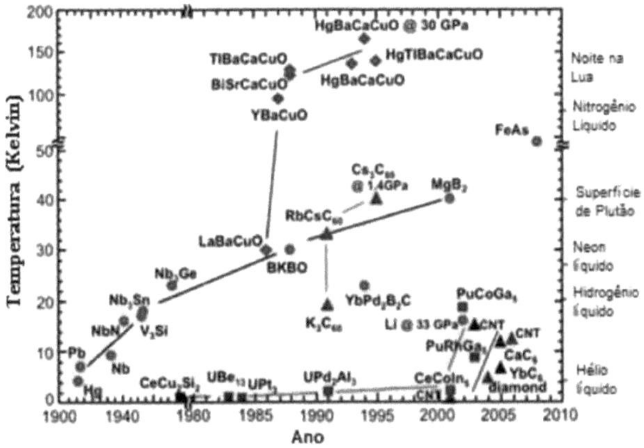

Source: Adapted from Dias (2011).

2.8 SUPERCONDUCTING SYSTEM Bi-sr-ca-cu-o

Like all superconducting oxides, BSCCO is a Type II high critical temperature (HTS) superconductor. This system was the first high-temperature superconductor not to contain a rare earth element

(MAEDA et al., 1988) and has a perovskite structure. The main representatives of this class of superconducting oxides are BSCCO and YBCO (POOLE JR. et al., 2007).

The perovskite structure has a chemical formula given by ABO_3, where A and B are metallic cations and O are non-metallic anions (GUARANY, 2008), as shown in Figure 11.

Figure 11. (a) Unit cell of an ABO perovskite structure3 and (b) structure visualized from the BO octahedrons .6

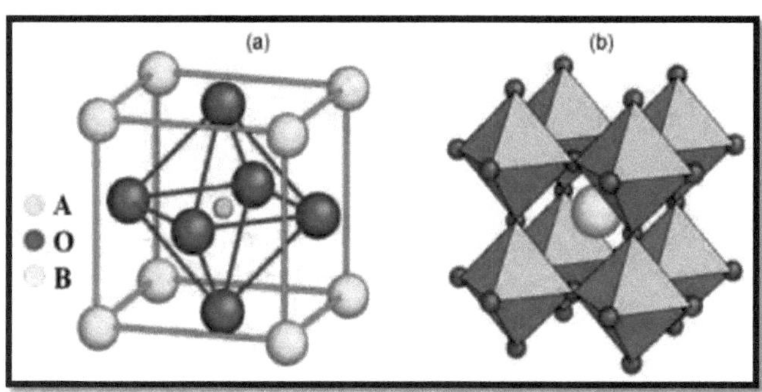

Source: Guarany (2008).

Bednorz and Müller (1986) in their first article on HTSs referred to their samples as "*metallic, oxygen-deficient... with mixed-valence copper perovskite compounds*". Subsequent work confirmed that the new superconductors did indeed possess these characteristics (BEDNORZ; MÜLLER, 1986; POOLE JR. et al., 2007).

Superconducting cuprates have the layer structure shown in Figure 12, with alternating bonding layers and conduction layers, where the CuO2 planes reside (POOLE JR. et al., 2007).

Figure 12. Schematic of the structure of superconducting cuprates, consisting of bonding and conduction layers.

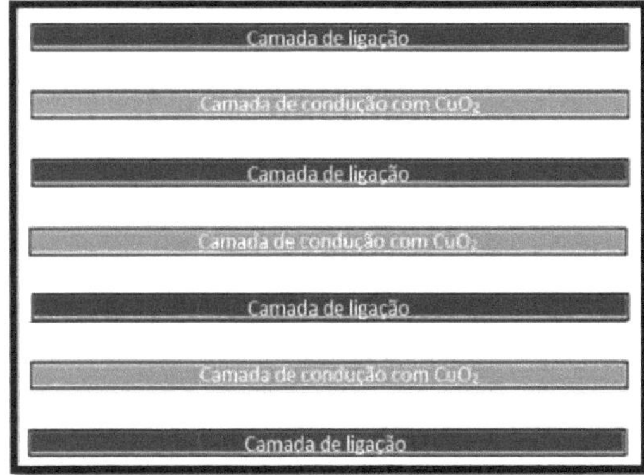

Source: Adapted from Poole Jr. et al. (2007).

In the conduction layer, the CuO pianos$_2$ are very close to planar geometry. Although in the normal state these materials behave like electrical insulators, the small amount of conduction they can exhibit is due to the electrons released by the Cu atoms and moving through the CuO pianos$_2$. In the superconducting state, it is believed that these same charge carriers are responsible for conduction in this state (POOLE JR. et al., 2007).

Our study focuses on BSCCO. This has several superconducting phases which are defined in relation to the number of layers of CuO_2 and Ca, as shown in Figure 13 (RODRIGUES, 2011).

Figure 13. (a) Schematic of the conduction and bonding layers in the BSCCO superconducting system, (b) the conduction layers for the different phases of the BSCCO system.

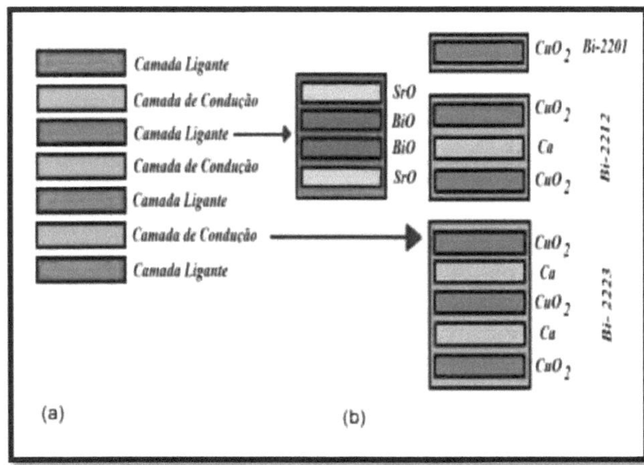

Source: Rodrigues (2011).

The BSCCO compound has four phases: 2201, 2212, 2223 and 2234, where the Tc's are, respectively, around 35K, 85K, 110K and 95K (POOLE JR. et al., 2007; RODRIGUES, 2011).

It should be noted that the bonding layers are also called charge reservoir layers, as they provide the charges for the carriers that form on the copper oxide planes (POOLE JR. et al., 2007).

The compound BSCCO (2212) is widely studied because of its extremely high planar anisotropy. The anisotropy of these materials is so high that in single crystals, where the crystallographic pianos are well defined, the resistivity in the orientation perpendicular to the CuO2 planes (parallel to the c axis) is up to three orders of magnitude greater than in the orientation parallel to the a or b axes. The crystal structure of this phase is shown in Figure 14 (HOFFMAN, 2010).

Figura 14. Crystal structure of the BSCCO 2212 phase.

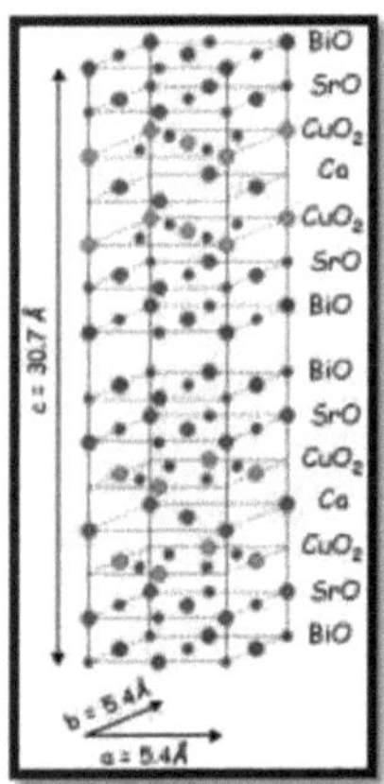

Source: Hoffman (2010).

Several studies have reported the use of BSCCO manufactured in the form of single crystals, long ribbons, wires, thin films and *bulks* (KATSUI; OHTSUKA, 1988; IYER et al., 2000; SALAMATI; KAMEL; AKHAVAN, 2004).

14.8.1 Phase Diagram

Phase diagrams can be used to obtain a wide variety of information, such as (1) the identification of compounds formed from finished elements, (2) the stability range, composition and temperature of a specific phase, (3) the conditions for synthesis, (4) the impurities that may be present with a phase prepared under specific conditions, (5) the region of homogeneity of solid solutions, (6) the polymorphism of a phase (7), temperature range and composition suitable for glass formation, (8) the melting temperature of a compound, (9) the composition for crystallization of a phase, (10) the effect of mixing elements on the melting point or stability of a phase (11), temperature effects on processing, and (12) the properties of the final products based on the assembly phase (WONG-NG, 2000).

Figure 15 shows, for a solid solution, the dependence of the stoichiometry of the composition Bi $_{.218}$ Sr_{3-y} Ca_y Cu O_{28+} X as a function of temperature. In this projection, the 2212 phase presents a single phase region, variable with *Sr, Ca, Bi* and *O*. This single phase region approaches a half-moon shape, with the greatest width of y at around 800°C. With increasing temperature, the extent of the monophase region is shifted towards Sr-rich compositions. The four-phase regions outside the single-phase area vary depending on the value of y in the formula. The high annealing temperature exhibited by the Ca rich 2212 phase leads to the precipitation of Ca_2 CuO_3 and a liquid, while the annealing of the Sr rich 2212 phase leads to the formation of Bi_2 Sr_3 Cu O_{288} , cuprates and a liquid. At temperatures above 870°C, the Sr-rich 2212 phase decomposes. The Sr:Ca ratio of the critical composition of phase 2212 has been determined to be about 2:1 (Bi2*.18Sr2CaCu2O8+d*). At the maximum melting temperature, the 2212 phase melts into 2201+cuprates+liquids (WONG-NG, 2000).

The 2212 solid solution was found to be in equilibrium with 10 phases at 830°C (WONG-NG, 2000). The equilibrium phases were *0x21* {[(Ca,Sr)$_2$ CuO_3], *x* is used to represent the concentration of the solid solution of the minor component}, 119x5 [(Bi,Pb)2.2Sr1.8-x CaxCuOz], 2110 [Bi16(Sr,Ca)14Oz], 014x24 [(Sr,Ca)14Cu24O41], 2310 [Bi2(Sr,Ca)4Oz], 4805 [Bi4Sr8Cu5Oz), 2201 [(Bi, Pb)2Sr2-xCaxCuOz], (Ca,Sr)O, CuO, and 0x11 [(Sr1-xCax)CuO2, rich in Ca]. This means that a small change in composition or temperature can lead to a large change in the equilibrium of the phases (WONG-NG, 2000).

Figura 15. Phase temperature-(Sr, Ca) stoichiometry for phase 2212 with composition Bi $_{.218}$ Sr_{3-y} Ca_y $Cu2O_{8+\delta}$.

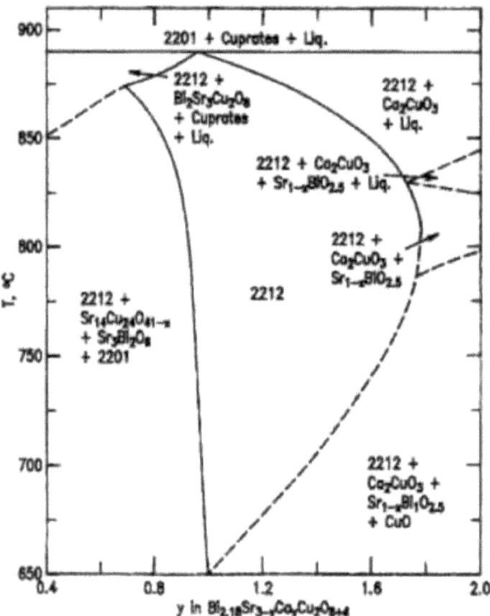

Source: Wong-NG (2000).

At the eutectic temperature, i.e. the lowest temperature (approximately 825°C) at which the 2212 phase is in equilibrium with the liquid phase, there is a liquid near the Sr-poor part of the Bi-Sr-Ca-Cu oxide system: 2110 + 119x5 + 2212 + CuO ÷L.

2.9The BSCCO SiNTESE

The main aim in producing superconducting materials is to obtain a material capable of carrying high current densities at the highest possible temperature. In order to achieve this goal with ceramic superconductors, there must be a change in the microstructure of the material to eliminate the two biggest obstacles to current flow: the "*Weak links*" between the superconducting grains, i.e. the connection between the grains and the homogeneity of these grains (MOTTA, 2009; RAO; NAGARAJAN; VIJAYARAGHAVEN, 1994). Alternative ways of improving the synthesis techniques for these materials are therefore being studied. Perovskite-type compounds, such as superconducting cuprates, have been synthesized by various methods. Conventional solid-state reaction methods can present problems of contamination and growth during the milling and calcination process. In recent years, wet chemical methods such as sol-gel, Pechini, precipitation and hydrothermalization methods have been used to overcome the limitations of the solid-state reaction method for synthesizing perovskites (RAO; NAGARAJAN; VIJAYARAGHAVEN, 1994; ZHANG et al., 2010).

The solid-state reaction is a simple synthesis method, but secondary phases cannot be removed, even through repeated grinding and heat treatment at high temperatures for some complex oxide multicomponents. In order to synthesize powders with high phase purity at low temperature, chemical methods have been investigated. The phase purity of ceramic powders can be increased by producing them via chemical routes because, in such a process, molecular homogeneity can be achieved and then fine powders can be obtained (RAO; NAGARAJAN; VIJAYARAGHAVEN, 1994; ZHANG et al., 2010).

2.9.1 Sol-Gel method

The Sol-Gel process consists of producing a solution and transforming it into a gel. A sol is therefore a dispersion of colloids and a gel is defined as a colloid or a polymeric solid, with one of its components in the liquid phase, both arranged in a dispersed manner. A concentrated sol of the oxides or hydroxides of the reactants is converted into a semi-rigid gel by removing the solvent. The dried gel is heated to a suitable temperature to obtain the product (BRINKER; SCHERER, 1989; MOTTA, 2009; RAO; NAGARAJAN; VIJAYARAGHAVEN, 1994; ZHANG et al., 2010), as shown in Figure 16 (CARDOSO, 2012). The sol-gel process is used to obtain a homogeneous mixture of cations. Most of the reactions in this process occur through hydrolysis and polycondensation. The microstructure, purity and physical properties of the products are controlled by the precursor reagents, pH, solvent and heat treatment (MOTTA, 2009; BRINKER; SCHERER, 1989). In the case of the sol-gel method, rigid agglomerations are formed through hydrogen bonds in the gel. The disadvantage of this method is the purity of the phase, which still needs to be improved (ZHANG et al., 2010).

Figure 16. Schematic diagram of the process and the formation of Sol-Gel.

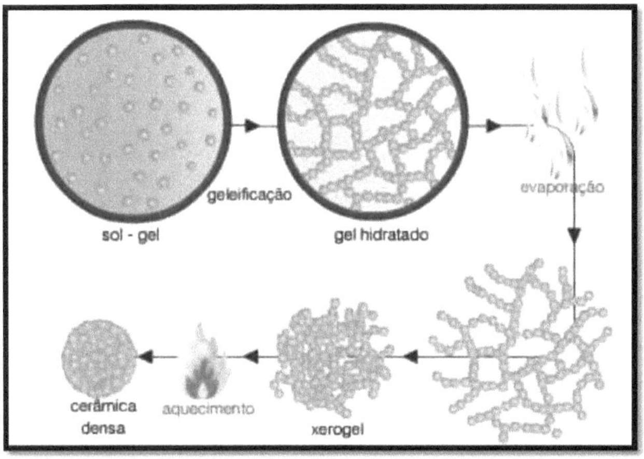

Source: Adapted from Cardoso (2012).

2.9.2 Pechini method

The Pechini Method, also known as the Polymer Precursor Method, was patented by Magio Pechini in 1967. In this work, the preparation of titanates and niobates is demonstrated, involving the ability of some a-hydroxycarboxylic acids (citric acid) to form chelates with most metal cations by adding polyalcohol (ethylene glycol). It is also essential to heat the solution to a moderate temperature so that it does not ignite, which is around 90°C. Thus, the polyesterification reaction takes place in which a polymer resin is formed in which the metal cations are incorporated, homogeneously distributed in the polymer chain (PECHINI, 1967), as shown in Figure 17. The immobilization of metal complexes in the rigid network of organic polymers reduces the segregation of specific metal ions, thus increasing the homogeneity of the composition (RAO; NAGARAJAN;

VIJAYARAGHAVEN, 1994; MOTTA, 2009; PECHINI, 1967). In order to eliminate the water and the organic part, it is necessary to carry out a suitable heat treatment starting at 300°C, in which the polymer chain or the bonds of the polymer bonds are broken (BRINKER; SCHERER, 1989; KAKIHANA, 1996). The citric acid/ethylene glycol ratio, which initiates gelling, is given by the mass ratio of 40/60. This ratio differs from the one used in the Pechini AG/EG Patent = 20/80. The 40/60 ratio is more commonly used because ethylene glycol, in excess, plays a role as a solvent to increase the initial solubility of the metal salts. The ethylene glycol must then be evaporated to obtain a polymeric gel, as described by Kakihana in 1996. The citric acid/metal ratio of 3:1 is very important (CHU; DUNN, 1987), as it defines the minimum amount of citric acid to coordinate all the metal ions.

Figure 17. Schematic diagram of the Pechini Method.

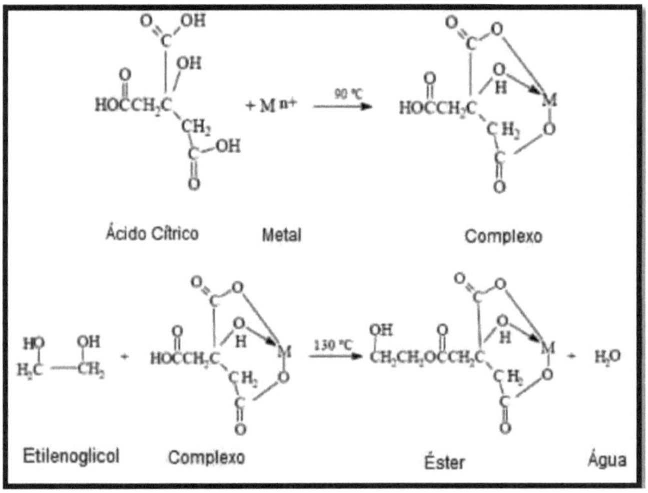

Source: Adapted from Pechini (1967).

The Pechini method was introduced due to the need to synthesize materials with high phase purity. Superconducting materials such as BSCCO (2212) are difficult to obtain because they have many phases and elements, which causes impurities to appear, thus lowering the critical temperature of the material (ZHANG et al., 2010).

2.9.2 Microwave-assisted hydrothermalization

The hydrothermal technique has been widely used in the synthesis of ceramic materials due to its advantages in the processing of nanostructured materials and, with this, it has greater appeal in technological applications (BYRAPA; ADSCHIRI, 2007). Byrapa and Yoshimura in 2001 defined hydrothermal as any heterogeneous chemical reaction in the presence of a solvent (aqueous or non-aqueous) above room temperature and at a pressure greater than 1 atm in a closed system.

The hydrothermalization process has many advantages such as obtaining high purity materials, controlled stoichiometry, high homogeneity and crystallinity, narrow particle size distribution and a lower sintering temperature of around 240°C (BYRAPA; YOSHIMURA, 2001; BYRAPA; ADSCHIRI, 2007). This processing method uses simple equipment with lower electricity consumption. However, synthesis times can reach up to a week (MOREIRA, 2010).

Microwaves were introduced to the conventional hydrothermalization method as a heat source in order to achieve a reduction in synthesis time and a rapid increase in temperature. This process became known as microwave-assisted hydrothermalization. Komarneni and colleagues, in 1992, were the first to introduce the synthesis of ceramics using the microwave-assisted hydrothermal method at a frequency of 2.45 GHz, the same used in domestic microwave ovens (MOREIRA, 2010; KOMARNENI; ROY; LI, 1992).

Microwave electromagnetic energy operates in the frequency range of 0.3 to 300 GHz (corresponding to wavelengths of 1 mm to 1 m). The common operating wavelength for domestic microwave ovens is 12.2 cm, which corresponds to a frequency of 2.45 GHz (MOREIRA, 2010; ZHU; HANG, 2012). In the microwave-assisted hydrothermalization process, the microwaves interact with the solution, and part of the electromagnetic energy is converted into thermal energy, which causes the material to heat up. As a result, the heat generated within the reaction becomes homogeneous due to the extremely rapid crystallization kinetics (MOREIRA, 2010; ZHU; HANG, 2012). The heat is generated from inside the material, in contrast to conventional heating, where heat is transferred from the outside to the inside through heat conduction.

Compared to the conventional way of heating, there are several advantages, such as fast heating rates, reduction in processing time, energy efficiency and all this provides more homogeneous microstructures leading to a better quality of the final product (ZHU; HANG, 2012).

In microwave heating, electromagnetic waves are produced in a magnetron and transmitted through a hollow metal tube (waveguide) to a resonance cavity where the materials are to be processed. The materials are heated due to their molecular friction caused by the alternating polarization (dipole) of molecules, which involves two main processes: dipole polarization and ionic conduction. Electromagnetic radiation is made up of oscillating fields and its interaction with the material's dipoles causes them to oscillate as well (AL-HARAHSHEH; KINGMAN, 2004; ZHU; HANG, 2012). Thus, different amounts of heat are produced through molecular friction and dielectric loss. As the rotational frequencies of many polar molecules in the liquid cause a delay in relation to the oscillation frequency of the electric field, resistive heating begins to be produced within the medium. This is called dielectric loss, which is the part of the energy of an alternating electric field in a dielectric medium that is converted into heat and dissipated in the sample. In the case of ionic conduction, the dissolved charged particles oscillate under the influence of microwave radiation and then collide with neighboring molecules, generating heat (AL-HARAHSHEH; KINGMAN, 2004; ZHU; HANG, 2012).

Several inorganic oxides, including CuO, absorb 2450 MHz radiation. In 1988, Baghurst and colleagues pioneered the use of a microwave oven to synthesize mixtures of metal oxides with superconducting properties. The heating takes place inside the copper oxides and is transferred from there to the surrounding area, allowing the crystallization temperature to be reached quickly and, consequently, resulting in a shorter synthesis time and a higher degree of purity of the material (BAGHURST; CHIPPINDALE; MINGOS, 1988; VOLANTI et al., 2007).

3 MATERIALS AND METHODS

3.ι MATERIALS

In this work, Bi:Sr:Ca:Cu:O compounds were prepared with the stoichiometry $Bi_2 Sr_2 CaCu_2 O_x$, i.e. the Bi-2212 phase, which was chosen for its chemical stability and absence of toxic elements. The compounds used to produce the samples were also varied, i.e. carbonates and nitrates were used. All the reagents used are shown in Table 1.

Table 1. Characteristics of the reagents used to synthesize the compounds.

Reagent	Chemical Formula	Manufacturer	Purity (%)
Bismuth Carbonate	$(BiO_2)CO_3$	Vetec	81
Strontium Carbonate	$SrCO_3$	Vetec	97
Calcium Carbonate	CaCO3	Vetec	98
Copper Carbonate	$CuCO_3 (OH)_2$	Vetec	56
Bismuth Nitrate	Bi5H9N4O22	Vetec	79
Strontium Nitrate	$Sr(NO_3)_2$	Vetec	99
Calcium Nitrate	$Ca(NO_3)_2$	Vetec	99
Copper Nitrate	$Cu(NO_3)_2$.3H2O	Dynamic	98
Citric acid	C6⅛O7	Dynamics	99,5
Nitric Acid	HNO_3	Dynamic	65
Ethylenediamine	C2⅛N2	Vetec	98
Ethylene glycol	C2H6O2	Vetec	98

Source: Author's research data.

The samples were then synthesized using the Pechini method (PM) and a combination of the Pechini method and the Microwave Assisted Hydrothermal method (MAH). A citric acid/metal molar ratio of 3/1 and a citric acid/ethylene glycol mass ratio of 40/60 were used.

3.2 METHODS

3.2.1 Preparation of precursors

The stoichiometries of the samples were calculated in order to obtain 5 g of material. To do this, the masses of the compounds were measured on an analytical balance; citric acid (C H O_{687}) was diluted in 32 mL distilled water using a magnetic stirrer and then $SrCO_3$ and $CaCO_3$ were added to the solution. Separately, (BiO_2)CO_3 and $CuCO_3$ $(OH)_2$ were diluted in a mixture of 5 mL of HNO_3 and 5 mL of distilled water. After diluting these components, $SrCO_3$ and $CaCO_3$ were added to the C H O_{687} solution and left to stir for 5 minutes.

Once this stage was complete, the pH was measured using a MsTecnopon mPA 210 pH meter, which showed a pH value of ≈ 1. Ethylenediamine (C H N_{282}) was then added slowly until the pH reached 8.251, which should be between 8 and 9. The total amount of C2H8N2 used was 23 mL and the final product had a dark purple color.

35.65 mL of ethylene glycol (c2H6o2) was then added to form the polymer. The solution was then stirred and heated to a temperature of 90°C for a period of 20 days. After this period of time, the solution had a coarse texture due to the evaporation of liquid and the material had a dark brown color.

3.2.2 Microwave-assisted hydrothermalization synthesis

To prepare the samples from MAH, the precursor solution was prepared in the same way as described above. However, after forming the polymer, the solution was handled inside a Teflon® autoclave and then fixed and sealed in the reaction chamber. The equipment used was a conventional microwave oven adapted for this use and belonging to the group at the National Institute for Materials Science and Technology in Nanotechnology at UNESP/FCT, shown in Figure 18.

Synthesis was carried out at a heating rate of 2°C/min and maintained at 140°C for 60 minutes. During the heating process, the pressure recorded was 3 bar. The solution was left to stand for 24 hours and the process was repeated once more. The solution contained in the reaction cell was cooled naturally until it reached room temperature. The resulting solution was a supernatant material and a dark brown viscous precipitate, which was separated and dried in an oven at 70°C for 5 days.

Figure 18. Domestic microwave oven adapted for synthesizing ceramic materials.

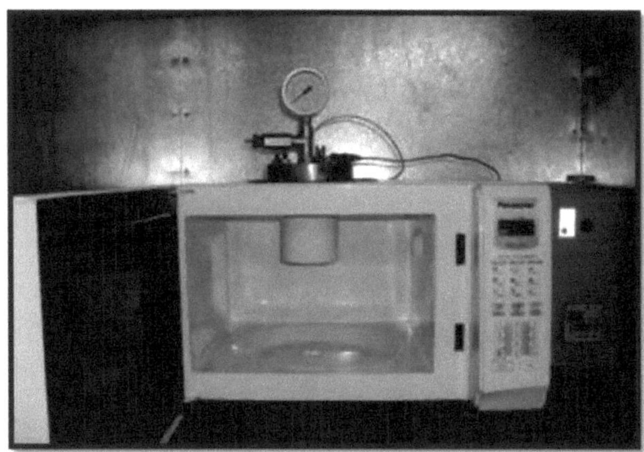

Source: Author's own elaboration.

The samples made with the nitrate regents were prepared in the same way as the carbonates, i.e. with the appropriate stoichiometry to obtain 5 g of material; C H O_{687} was diluted in 32 mL of distilled water using a magnetic stirrer, to which Sr(NO $)_{32}$ and Ca(NO $)_{32}$ were added. Bi H N_{59422} Oand Cu(NO $)_{32}$ .3H_2 O were diluted separately in 5 mL of distilled water plus 5 mL of HNO_3 and then this mixture was added to the previous solution and the system was kept stirring for 5 minutes. The pH of this solution was approximately 1. Ethylenediamine ($C_2H_8N_2$) was then added until a pH = 8.16 was reached. This required 26 mL of C H N_{282} and the final product had a dark purple color.

To form the polymer, 35.65 mL of ethylene glycol (C H O_{262}) was added to the solution and transferred to the Teflon® flask, duly sealed and fixed in the reaction cell. The synthesis was carried out at the same heating rate as for the carbonate sample and the pressure recorded in the system was also 3 bar. The solution was left to stand for 24 hours and the process was repeated once more. The result was a supernatant and a dark brown precipitate which was separated and left in an oven for 5 days at 70°C.

The samples were named according to the differences in their syntheses, as shown in Table 2.

Table 2. Sample nomenclature.

Samples	**Synthesis**
Ref	Reference sample, MP with carbonate reagents.
CaMAH	MP coupled to MAH with carbonate reagents.

NiMAH	MP coupled to MAH with Nitrate reagents.

Source: Author's research data.

3.2.3 Heat treatment

Appropriate heat treatment is important for the formation of the material's superconducting phase. To eliminate the organic part, the solution is subjected to a treatment of up to 400°C, so that above this temperature the carbon CO_2 is eliminated and the oxides are formed (PENG et al., 1998). For the phase to form, sintering treatment is required, which, for Bi-2212, takes place above 800°C (PENG et al., 1998). During the sintering stage, the particles fuse as shown in Figure 19.1 and, along the contact regions between them, there is deformation and the appearance of a grain contour, shown in Figure 19.2. In this process, each interstice between particles becomes a pore, Figure 19.3. As the sintering process progresses, the pores become smaller and acquire spherical shapes, and the grains grow (CALLISTER JUNIOR, 2008). The process described is shown in Figure 19.4.

Figure 19. Grain growth mechanism after sintering.

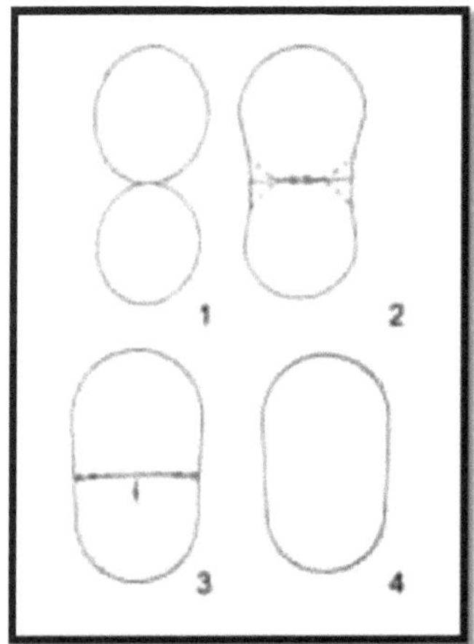

Source: Adapted from Motta (2009).

The appropriate heat treatments were applied to all the samples using an EDG muffle furnace, model EDGCON 3P, to remove the organic matter and water. Calcination was carried out at 200°C/1h and 400°C/2h and, in the sintering process, the powders obtained were subjected to heat treatments at 835°C, changing the treatment time between 2h and 6h, with a heating rate of 2°C/min. These powders were then characterized by X-ray diffraction (XRD) and scanning electron microscopy (SEM).

To measure the transport properties, tablets with 1.2 g of powder were prepared using a Bovenau model P15ST uniaxial hydraulic press at 10 tons. The pellets were then heat-treated at 845°C/24h using an EDG FT HV series tube furnace. After this treatment, X-ray diffraction (XRD), scanning electron microscopy (SEM), energy dispersive X-ray spectroscopy (EDS), electrical measurements and magnetic measurements were carried out.

To check the influence of the heat treatment on the samples and reduce the segregated phases still present, a new treatment was carried out at 845°C/72h using a tube furnace. After this treatment, X-ray diffraction (XRD), energy dispersive X-ray spectroscopy (EDS), scanning electron microscopy (SEM) and electrical measurements were carried out again.

3.2.4 Characterizations

3.2.4.1 X-ray diffraction (XRD)

The crystalline phases of the samples were analyzed by X-ray diffraction using a Shimatzu XRD-6000 diffractometer from the DFQ Polymer Group at UNESP/FEIS. The measurements were made at a speed of 1°/min, with a reading interval of 0.02° and a 2θ scan between 4° and 60°, using a wavelength λ = 1.542A with a CuKa radiation filter.

The XRD technique is used to determine crystal structure. In it, a beam of X-rays is directed at a crystalline material and can undergo diffraction. The rays are diffracted in such a way as to generate patterns of constructive and destructive interference. The result of constructive interference is a pattern of peaks that are related to the spacing between the planes of the material's crystal lattice and the same order of magnitude as the X-ray wave is obtained according to Bragg's law (CALLISTER JUNIOR, 2008).

$$n\lambda = 2d\ sen\theta \tag{16}$$

where *n is* an integer, *λ* is the wavelength of the incident radiation, *d is* the distance between atomic planes and θ is the angle of incidence with respect to the crystal plane.

The diffractograms were analyzed using the program X'Pert HighScore, by Philips Analytical B.V., 2001, and also the database "*Inorganic Crystal Structure Database (ICSD)*".

3.2.4.2 Scanning Electron Microscopy (SEM)

The morphology and microstructure analyses were carried out using a ZEISS EVO LS15 scanning electron microscope. This device emits a focused beam of electrons onto the surface of the sample, which is reflected and collected, and then displayed at the same scan rate on a cathode ray tube. The captured image shows the characteristics of the sample surface. The surface must be conductive in order for the electrons incident on the sample to flow. If the sample is not conductive, a metallic coating must be deposited on its surface (CALLISTER JUNIOR, 2008). Magnifications ranging from 10 to over 50000 diameters are possible.

The preparation of the powdered samples for the SEM consisted of dissolving them in isopropyl alcohol ($C\ H_{38}\ O$) and putting them through an ultrasonic cleaner (Ultrasonic Cleanear T-14, Odontobrâs) for 5 minutes. The solution with the sample particles was poured onto a sample holder, dried and then taken to the SEM.

The tablet-shaped samples were broken and fixed in the sample holders with the damaged side facing upwards. Silver paint was used for fixing. The face was then analyzed and images taken at 5000x magnification.

Images were also taken using the backscattering electron (BSE) detector. Image recording uses a second high-resolution cathode ray tube and a conventional coupled camera system (DEDAVID; GOMES; MACHADO, 2007).

The image generated by these electrons provides different information in relation to the contrast they present: a topographic image (contrast depending on the relief) and also a compositional image (contrast depending on the atomic number of the elements present in the sample through shades of gray) (DEDAVID; GOMES; MACHADO, 2007). The images were taken at 1000 times magnification.

3.2.4.3 Energy Dispersive X-ray Spectroscopy (EDS)

For the X-ray energy dispersive spectroscopy (EDS) analysis, the same equipment as the SEM was used, in which there is an X-ray detector. This technique consists of measuring the characteristic X-rays emitted in a certain region of the sample bombarded by an electron beam. The characteristic X-ray lines are specific to the atomic number of the sample and their wavelength or energy can be used to identify the element that is emitting the radiation (DEDAVID; GOMES; MACHADO, 2007). The analyses were carried out in different regions.

3.2.4.4 Electrical measurements

Electrical measurements were made using the four-point DC method, as shown in the diagram in Figure 20. This technique provides information on the critical temperature, T_c, of the material, which makes it possible to identify the superconducting phases present and parameters for obtaining the critical current density, J_c, (POOLE JR. et al., 2007).

Figure 20. Measurement scheme using the four-point dc method.

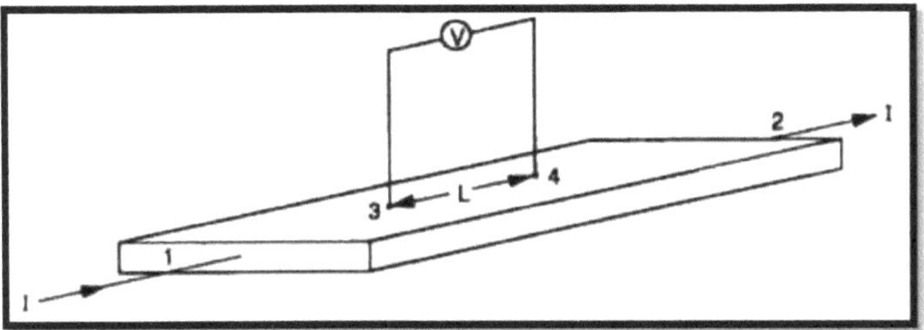

Source: Poole Jr. et al. (2007).

Two wires carry a known current in and out of the ends of the sample and two other conductors, separated by a distance *L*, measure the voltage drop at points closer to the center, where the current approaches steady-state flow. The resistance *R* is measured between points 3 and 4, as shown in Figure 20, and the resistivity is obtained using Ohm's law (POOLE JR. et al., 2007).

$$R = \rho \frac{L}{A} \quad (17)$$

where *A* is the cross-sectional area.

From the electrical measurements, we obtained the electrical resistance as a function of temperature using a Keithley 228A current source; a Keithley 2000 multimeter, a Keithley 2182 nanovoltmeter and a "Dewar" in which cryogenic liquid was placed, i.e. liquid nitrogen, the temperature of which is 77K. The measurements were carried out at a temperature variation rate of 1K/min and an electric current of 1 mA.

In these measurements, cut wafers were used and fixed in the sample holders, made of printed circuit board, and then the electrical contacts were made.

3.2.4.5 Magnetic measurements

The magnetic characterization of the samples was carried out using DC and AC magnetometry techniques.

Thus, magnetization measurements were made as a function of temperature, MxT, following the *ZFC* and *FC* processes, which were described in Section 2.6. From these measurements, T_c and the irreversibility temperature T_{ir}, associated with the fusion of the vortex network in the material, were obtained. The measurements were carried out on a Quantum Design magnetometer, a PPMS model 6000 from the Superconductivity and Magnetism Group at the Federal University of Sao Carlos,

UFSCar. This system works with liquid helium between 1.8 K and 350 K. In particular, the MxT curves were obtained for applied fields of 10 and 100 Oe.

In the AC susceptibility measurements, an alternating magnetic field was applied to the sample. In such a system, the response of the sample to this excitation is detected by a pair of coils that form a second-order gradiometer. The measurements were carried out using a frequency of 1 kHz and amplitudes of 0.01 and 1.00 Oe.

The AC susceptibility is given by a real component (χ') and an imaginary component (χ').

$$\chi = \chi' + i\chi'' \qquad (18).$$

The real component describes the shielding currents in the sample and the imaginary component is associated with dissipative processes in the materials (CARDOSO, 2001).

4 RESULTS AND DISCUSSION

4.1 STRUCTURAL IDENTIFICATION OF THE PHASES

From the XRD of the sample Ref pelletized and heat-treated at 845°C for 24 hours, when comparing it with the reference charts from the X'Pert HighScore program, it can be seen that the 2212 phase was formed, as shown in Figure 21. Several charts of the same phase were used because, as oxygen is not controlled in the synthesis process, it can vary. In the diffractogram of the Ref sample subjected to heat treatment at 845°C for 72 hours, shown in Figure 22, it can be seen that, even with the increase in heat treatment time, there was no major change and the predominant phase continued to be 2212. A comparison of all the treatments, from the first treatment to form the powder to the last treatment of the pelletized sample, is shown in Figure 23. Note that the 2212 phase is found from the first treatment, however, with increasing treatment time, some peaks of segregated material are no longer present in the pelletized samples.

Figura 21. Diffractogram of the Ref sample heat-treated at 845°C/24h. In the legend, the number in brackets refers to the ICDD-JCPDS file.

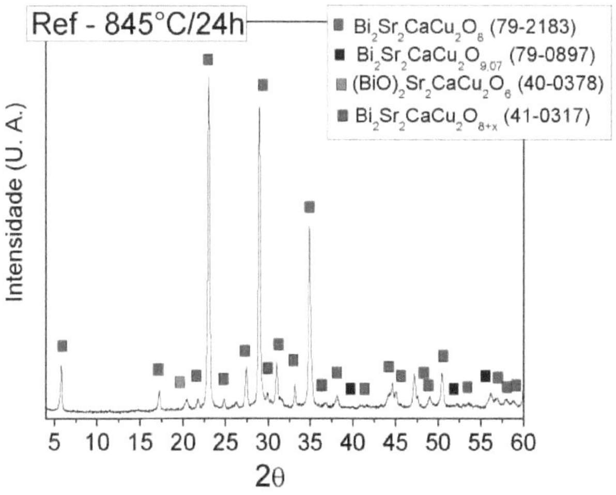

Source: Author's own elaboration.

Figura 22. Diffractogram of sample Ref heat-treated at 845°C/72h. In the legend, the number in brackets refers to the ICDD-JCPDS file.

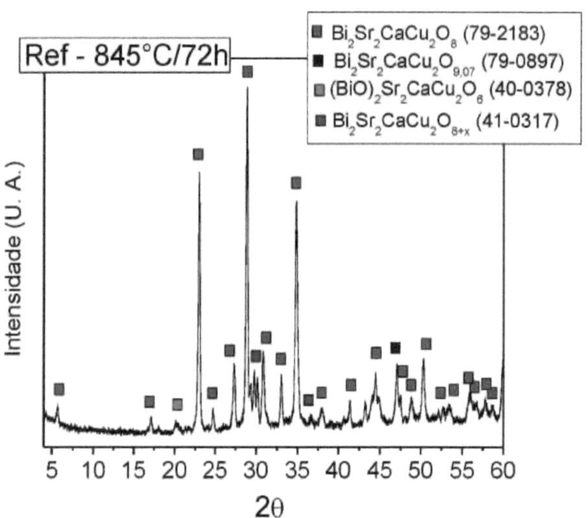

Source: Author's own elaboration.

Figura 23. Diffractogram of the Ref sample with all the heat treatments used.

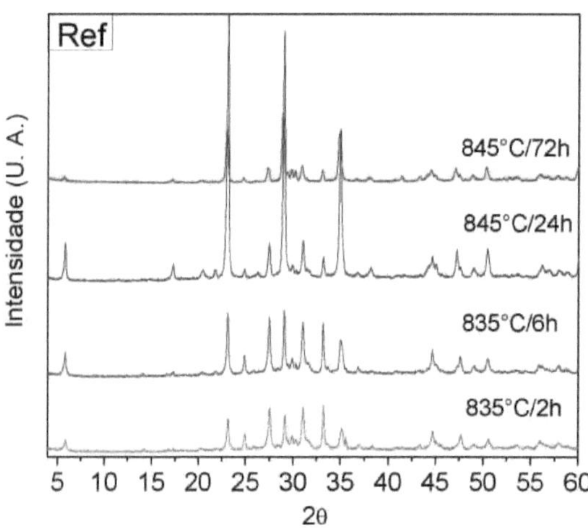

Source: Author's own elaboration.

For the CaMAH sample heat-treated at 845°C/24h, the diffractogram shows a predominance of the 2212 phase but also some peaks indicating the 2201 phase, as shown in Figure 24. When the sample is re-treated at 845°C/72h, we can see in Figure 25 that the amount of the 2201 phase decreases slightly and the 2212 phase peaks become larger. Figure 26 shows a comparison between all the heat treatments. We can see that there is a significant reduction in segregated phases.

Figure 24. Diffractogram of the CaMAH sample heat-treated at 845°C/24h. In the legend, the number in brackets refers to the ICDD-JCPDS file.

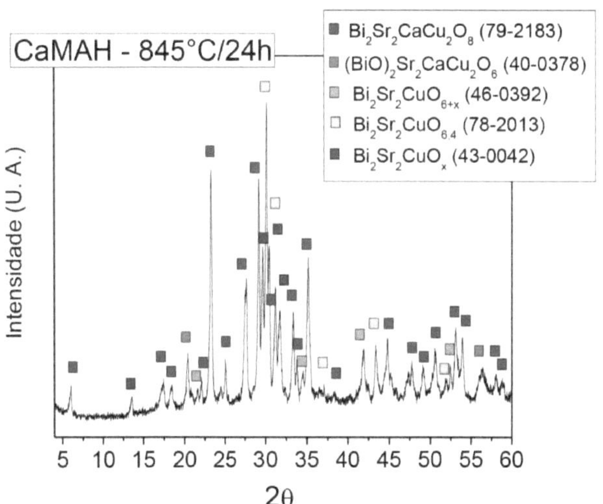

Source: Author's own elaboration.

Figure 25. Diffractogram of the CaMAH sample heat-treated at 845°C/72h. In the legend, the number in brackets refers to the ICDD-JCPDS file.

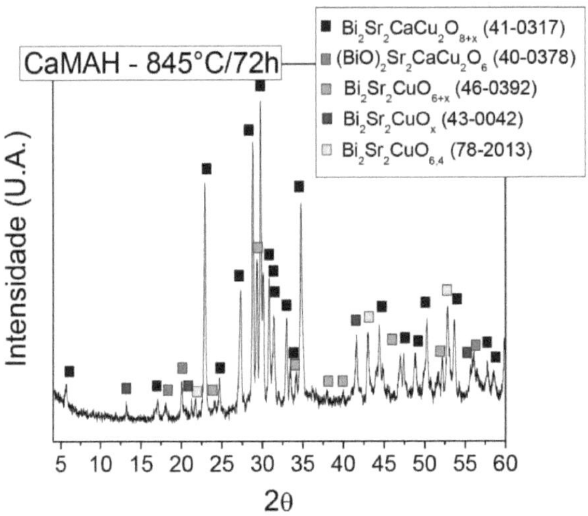

Source: Author's own elaboration.

Figura 26. Diffractogram of the CaMAH sample with all the heat treatments used.

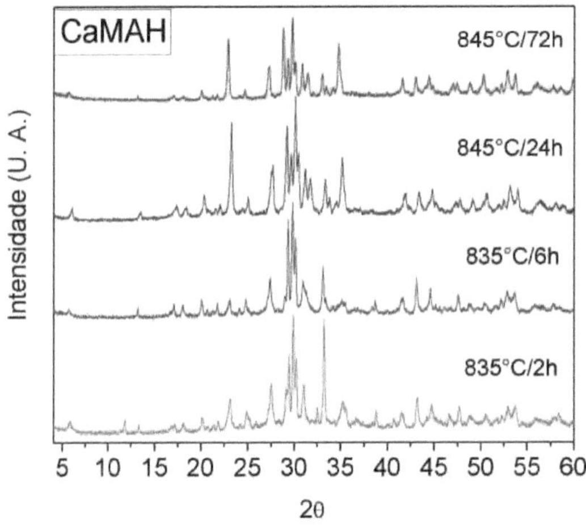

Source: Author's own elaboration.

For the NiMAH sample heat-treated at 845°C/24h, it can be seen from the diffractogram in Figure 27 that, like the other samples, the predominant phase is 2212, although it also contains phase 2201. The diffractogram also shows the presence of a peak for the SrCuO2 compound and another for the Bi2Sr3O6 compound. It can therefore be concluded that other compounds were formed which were not expected.

Figure 28 shows the diffractogram of the NiMAH sample treated at 845°C/72h. Note that as the heat treatment time increased, the peaks related to the 2212 phase increased and the segregated phase of the SrCuO2 compound was eliminated. In the diffractograms in Figure 29, we can see that, as the heat treatment time increases, the segregated phases decrease.

Figura 27. Diffractogram of the NiMAH sample heat-treated at 845°C/24h. In the legend, the number in brackets refers to the ICDD-JCPDS file.

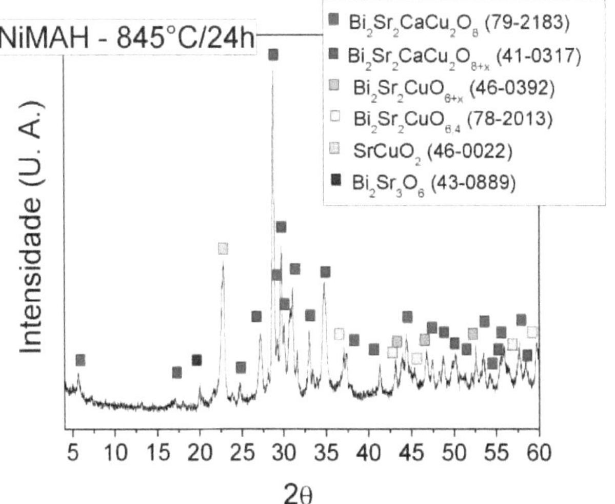

Source: Author's own elaboration.

Figura 28. Diffractogram of the NiMAH sample heat-treated at 845°C/72h. In the legend, the number in brackets refers to the ICDD-JCPDS file.

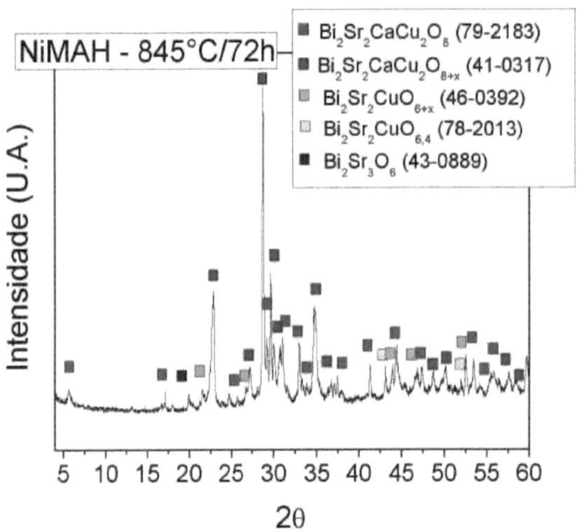

Source: Author's own elaboration.

Figura 29. Diffractogram of the NiMAH sample with all the heat treatments used.

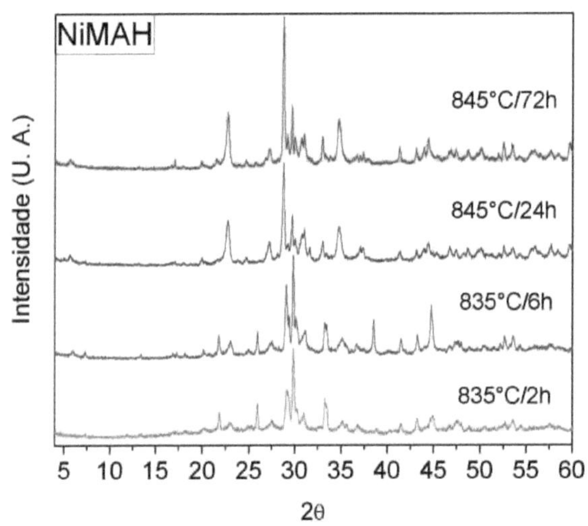

Source: Author's own elaboration.

1.2 SCANNING ELECTRON MICROSCOPY (MEVJ

Scanning electron microscopy images showed the morphology of the samples. Figure 30 shows the

SEM images of the samples (a) Ref, (b) CaMAH and (c) NiMAH, in powder form, heat-treated at 835°C/6h. Grains are observed in the form of plates, which are characteristic of BSCCO (MAJEWSKI, 2000). It can be seen that the Ref sample (Figure 30(a)), even in powder form, already has large plates. In the CaMAH sample, Figure 30(b), there are overlapping plates of various sizes very close together with the presence of several small grains. In Figure 30(c), the NiMAH sample, small plates can be seen.

Figure 30. Micrograph of the powders of the samples heat-treated at 835°C/6h. (a) Ref sample, (b) CaMAH sample and (c) NiMAH sample.

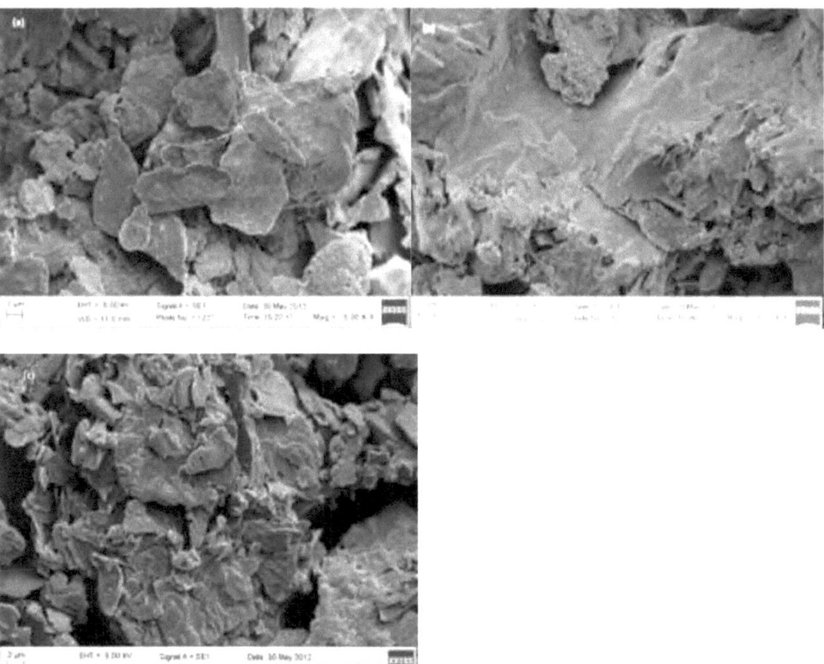

Source: Author's own elaboration.

SEM images were also taken of the pastillated and heat-treated samples at 845°C/24h, which are shown in Figure 31. It can be seen that, apart from the plaques, the samples had grains of different sizes. In Figure 31(a) it can be seen that the Ref sample has plaques and small grains whose dimensions are, on average, 1 μm. Figure 31(b) and (c) shows the formation of larger plaques and grains. This is due to the use of MAH, whose observed morphology may be associated with the way heat is transferred to the material, causing an increase in the kinetics of crystallization and grain growth. For the CaMAH sample, the grains are approximately 6 μm in size and for NiMAH, approximately 4 μm. As a note, in this work two nomenclatures will be used to distinguish the grains of the samples in terms of their shape, i.e. the grains in the shape of plates will be referred to simply as *plates* and those with a spherical shape will be referred to as *grains*.

It is important to note that the superconducting properties of these materials are highly dependent on the size and homogeneity of the grains, as well as the connectivity between them (MOTTA, 2009; POOLE JR. et al., 2007).

Figure 31. Pelletized samples heat-treated at 845°C/24h. (a) Ref sample, (b) CaMAH sample and (c) NiMAH sample with grain sizes.

Source: Author's own elaboration.

Images were also taken using the backscattered electron detector (BSE). Figure 32 shows the images of the Ref, CaMAH and NiMAH samples treated at 845°C/24h. In Figure 32(a), it can be seen that the gray tone is homogeneous, with small darker regions, which indicate the presence of pores. Figures 32(b) and (c) show regions with different shades of gray (darker gray), indicating that the atomic number of the material's elements are different.

Figure 32. Micrographs of the pelletized samples heat-treated at 845°C/24h using the BSE detector:

(a) Ref sample, (b) CaMAH sample and (c) NiMAH sample. The dark regions are due to the presence of pores and the contrasting gray regions indicate that they have a different atomic number.

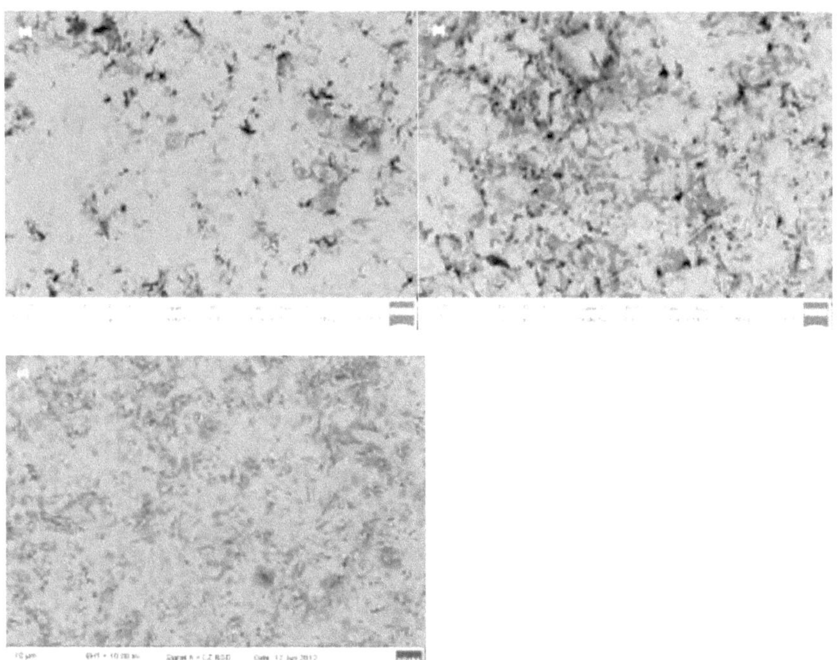

Source: Author's own elaboration.

After the samples had been heat-treated at 845°C/72h to eliminate the segregated phases, the SEM analysis was carried out again. The images obtained are shown in Figure 36. For the Ref sample, shown in Figure 33(a), it can be seen that the number of small grains has decreased and the number of plates has increased. For the CaMAH and NiMAH samples, shown in Figure 33(b) and (c), there is an increase in the size of the plates, but there are still a lot of grains that were not melted in the heat treatment process. The NiMAH sample also shows a change in the morphology of some of the plates, which are now rod-shaped.

Figure 33. Micrograph of the pelletized samples heat-treated at 845°C/72h. (a) Ref sample, (b) CaMAH sample and (c) NiMAH sample.

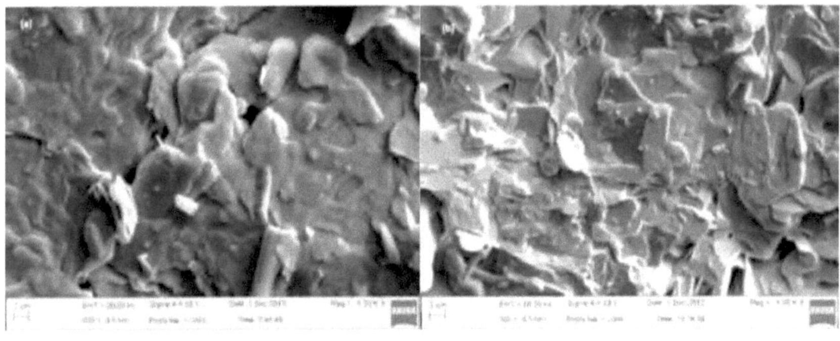

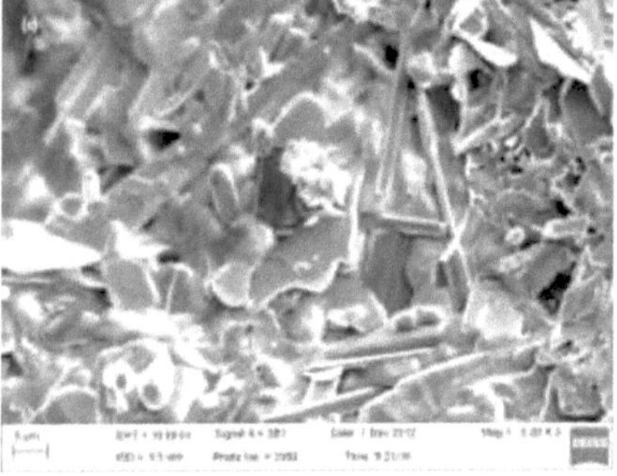

Source: Author's own elaboration.

The images of the samples in Figure 34, however, obtained by BSE, are shown in Figure 34. In Figure 34(a), referring to the Ref sample, it can be seen that there has been no change in the shade of gray, i.e. there is homogeneity in the color and, consequently, in the atomic number of the elements in the material. In Figure 34(b), referring to the CaMAH sample, it can be seen that where there are grains, the gray tone is darker, but in regions where there are plates, the colors are lighter. There was also a slight decrease in agglomeration. In Figure 34(c), referring to the NiMAH sample, it can be seen that there has been a reduction in the number of grains due to the lower contrast shown in the image.

Figure 34. Micrographs of the pelletized samples heat-treated at 845°C/72h using the BSE detector. (a) Ref sample, (b) CaMAH sample and (c) NiMAH sample. The dark regions are due to the presence of pores and the contrasting gray regions indicate that they have different atomic numbers.

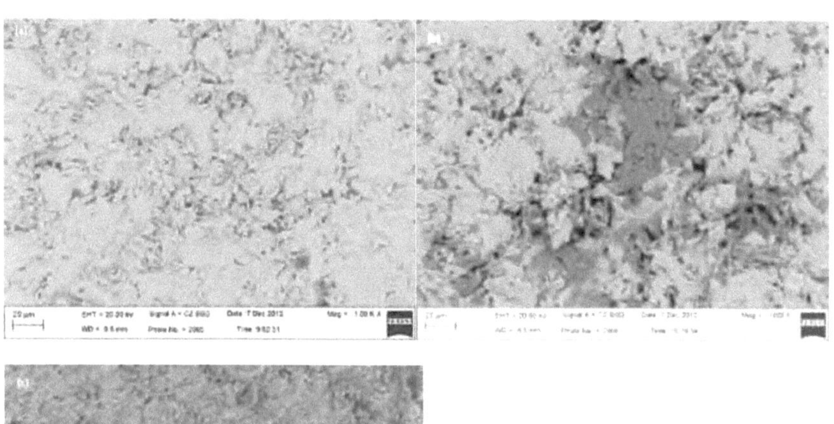

Source: Author's own elaboration.

It is possible to make an association between the XRD peaks and the morphological evolution of the samples shown by the SEM.

It can be seen that when the plates increase, the diffraction peaks decrease for the Ref and CaMAH samples (with increasing treatment time). For the NiMAH sample, as the presence of plates decreases, the rods of the diffraction peaks increase. Therefore, the decrease in the intensity of the XRD peaks is associated with grain growth for the Ref and CaMAH samples for the 2212 phase. And the increase in XRD peak intensities for the NiMAH sample may be associated with a decrease in grain size and/or morphology (rods).

1.3 Energy dispersive X-ray spectroscopy (EDS)

EDS analyses were carried out on all the samples and in three different regions, as shown in the SEM images in Figures 35 to 52 in which the BSE detector was used. Different shades of gray can be seen that differentiate the grain and plaque regions. Therefore, the analysis was carried out in three different regions, one focusing on the grains, another on the plaques and a third covering a large area of the sample in order to obtain the average composition of the samples. The $Bi_2Sr_2CaCu_2O_x$ stoichiometry was used for normalization, in which case no oxygen was added because there was no control over the process.

Tables 3, 4 and 5 show the composition of the Ref sample treated at 845°C/24h. Figure 35 shows the largest region used in the EDS analysis for this sample. Table 3 shows the composition of Figure 35, i.e. $Bi_{2.16}Sr_{1.93}Ca_{0.93}Cu_{2.03}O_x$. Figure 36 shows the region of the grains indicated by an arrow and its composition, which is shown in Table 4, is $Bi2_{,19}$ Sr $_{,19}$ 3Ca $_{,09}$ 5Cu $_{,19}$ $2O_x$. Figure 37 shows the region of the plates whose data is contained in Table 5, where it can be seen that the composition is $Bi_{2.09}Sr_{1.88}Ca_{0.93}Cu_{2.09}O_x$. As can be seen, all the regions analyzed have similar stoichiometries which are very close to the desired stoichiometry, i.e. $Bi_2Sr_2CaCu_2O_x$.

Figura 35. Image of a larger area analyzed by EDS for sample Ref - 845°C/24h.

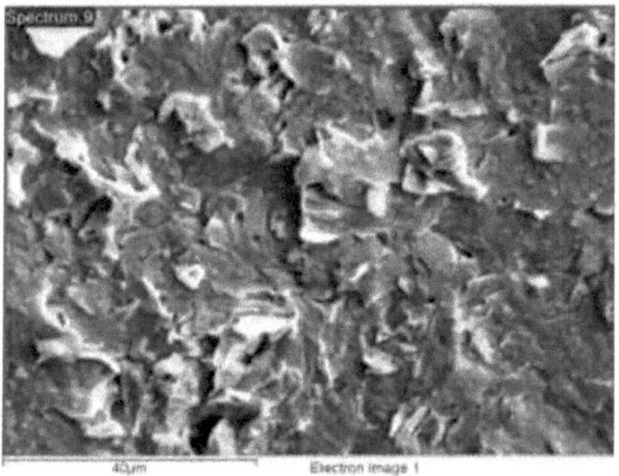

Source: Author's own elaboration.

Table 3. EDS results for sample Ref - 845°C/24h in the region of the total area with grains and plates.

Element	Atomic %	Nominal	EDS
Bi	30,90	2	2,16
Mr	27,66	2	1,93
Ca	13,35	1	0,93
Cu	28,08	2	2,03

Source: Author's research data.

Figura 36. Image of a grainy area indicated with an arrow, taken by EDS analysis for sample Ref - 845°C/24h.

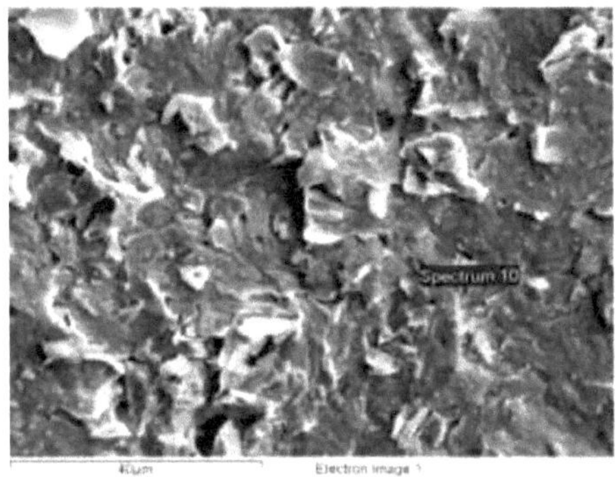

Source: Author's own elaboration.

Table 4: EDS results for sample Ref - 845°C/24h in the grain region.

Element	Atomic %	Nominal	EDS
Bi	31,26	2	2,19
Mr	27,64	2	1,93
Ca	13,60	1	0,95
Cu	27,64	2	1,92

Source: Author's research data.

Figura 37. Image of the region where the plaques were analyzed by EDS for sample Ref - 845°C/24h.

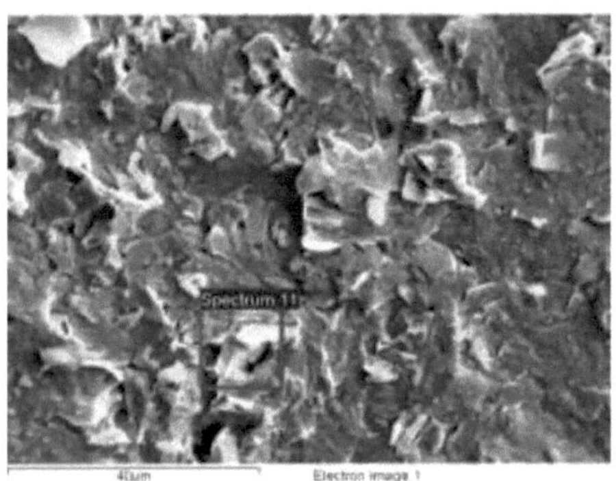

Source: Author's own elaboration.

Table 5 EDS results for sample Ref - 845°C/24h in the region containing plaques.

Element	Atomic %	Nominal	EDS
Bi	29,90	2	2,09
Mr	26,93	2	1,88
Ca	13,21	1	0,92
Cu	29,95	2	2,09

Source: Author's research data.

Tables 6, 7 and 8 contain the compositions of the CaMAH sample heat-treated at 845°C/24h, Figure 38 shows the largest region used in the EDS analysis, the data obtained is in Table 6, which is $Bi_{1.93}Sr_{2.05}Ca_{0.94}Cu_{2.08}O_x$. Figure 39 shows the region of grains indicated by an arrow, whose data in Table 7 is $Bi_{2.99}Sr_{2.41}Ca_{1.10}Cu_{0.48}O_x$ and Figure 40 shows the region containing plates, whose data in Table 8 is $Bi_{2.10}Sr_{1.91}Ca_{0.90}Cu_{2.08}O_x$. These results prove what was observed in the SEM with the BSE detector, where all the grain regions have different stoichiometries from the plate region. Once the entire region is observed, the desired stoichiometry is obtained. Knowing that EDS is a qualitative analysis.

Figura 38. EDS image of a larger area with grains and plaques for the sample CaMAH - 845°C/24h.

Source: Author's own elaboration.

Table 6. EDS results for the CaMAH - 845°C⁄24h sample in the larger area region.

which contains grains and plates.

Element	Atomic %	Nominal	EDS
Bi	37,61	2	1,93
Mr	29,21	2	2,05
Ca	13,43	1	0,94
Cu	29,74	2	2,08

Source: Author's research data.

Figura 39. Image of the grain region indicated by the arrow, taken by EDS analysis for the CaMAH - 845°C/24h sample.

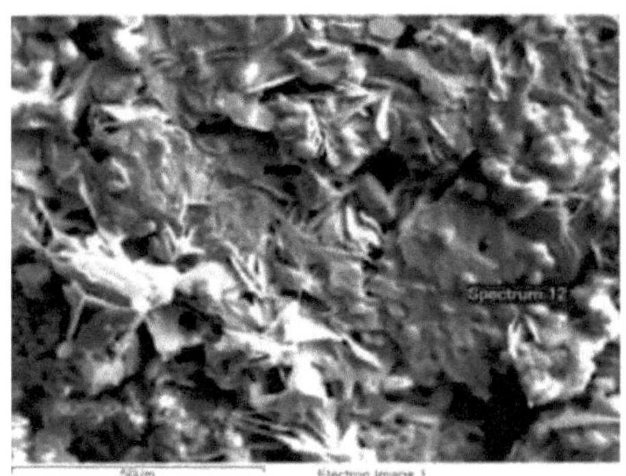

Source: Author's own elaboration.

Table 7. EDS results for the CaMAH - 845°C/24h sample in the area containing grains.

Element	Atomic %	Nominal	EDS
Bi	42,74	2	2,99
Mr	34,47	2	2,41
Ca	15,79	1	1,10
Cu	6,98	2	0,48

Source: Author's research data.

Figura 40. Image of the plaque region, indicated by an arrow, taken by EDS analysis for the CaMAH - 845°C/24h sample.

Source: Author's own elaboration.

Table 8. EDS results for the CaMAH - 845°C/24h sample in the plate region.

Element	Atomic %	Nominal	EDS
Bi	30,09	2	2,10
Mr	29,73	2	1,91
Ca	12,82	1	0,90
Cu	29,63	2	2,08

Source: Author's research data.

Tables 9, 10 and 11 contain the compositions of the NiMAH sample heat-treated at 845°C/24h. For Figure 41, the region of the larger area used in the EDS analysis was analyzed. As with the other samples, the data obtained by EDS is contained in Table 9, thus showing the compound Bi2.18Sr2.32Ca0.66Cu1.830x. Figure 42 shows the region of grains indicated by the arrow, whose data are shown in Table 10 for the Bi0.88Sr3.41Ca1.04Cu2.00Ox phase and Figure 43 shows the region of plates, indicated by the arrow, whose data are shown in Table 11 for the compound

Bi2,07Sr2,84Ca0,44Cu1,61Ox. Again proving what was observed in the SEM with the BSE detector, all the grain regions have different stoichiometries from the desired one, the plates have stoichiometries close to the quantitatively expected one, but when the whole region is observed, the desired stoichiometry is obtained.

Figura 41. Image of the area analyzed by EDS for the NiMAH -845°C/24h sample.

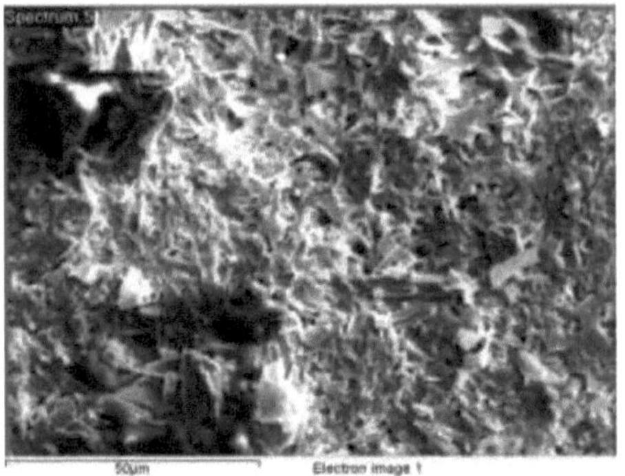

Source: Author's own elaboration.

Table 9. EDS results for the NiMAH - 845°C/24h sample in the larger area containing plates and grains.

Element	Atomic %	Nominal	EDS
Bi	31,19	2	2,18
Mr	33,18	2	2,32
Ca	9,45	1	0,66
Cu	26,16	2	1,83

Source: Author's research data.

Figura 42. EDS image of a grain area for the NiMAH sample - 845°C/24h.

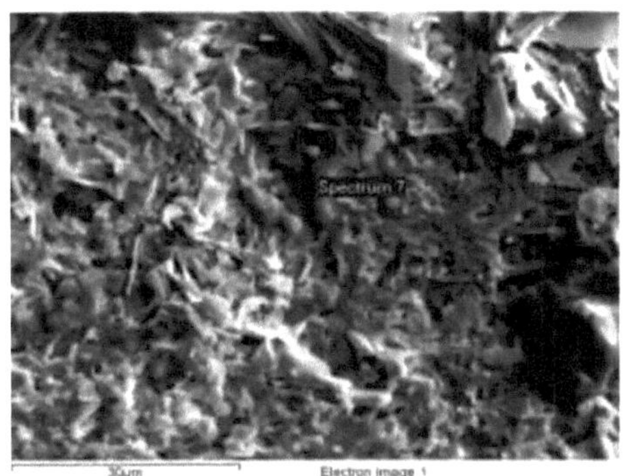

Source: Author's own elaboration.

Table 10. EDS results for the NiMAH - 845°C/24h sample in the grain area region.

Element	Atomic %	Nominal	EDS
Bi	11,79	2	0,82
Mr	48,79	2	3,41
Ca	14,94	1	1,04
Cu	24,46	2	2,00

Source: Author's research data.

Figura 43. Image of the region of a plaque area made by EDS analysis for the NiMAH - 845°C/24h sample.

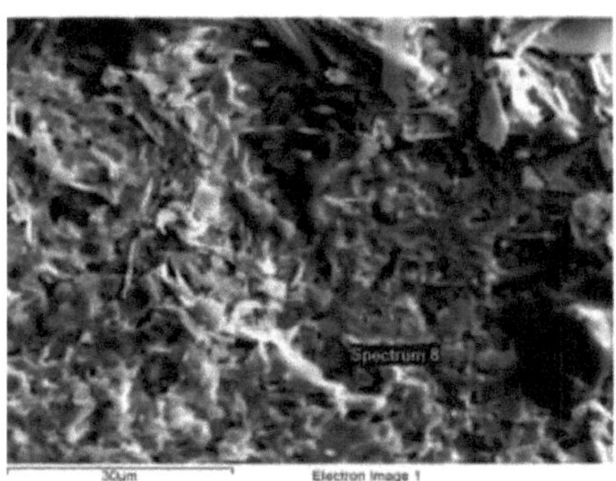

Source: Author's own elaboration.

Table 11. EDS results for the NiMAH - 845°C/24h sample in the plaque area region.

Element	Atomic %	Nominal	EDS
Bi	29,61	2	2,07
Mr	40,66	2	2,84
Ca	6,33	1	0,44
Cu	23,06	2	1,61

Source: Author's research data.

Analyses were also carried out on samples heat-treated at 845°C/72h, for sample Ref. These data are shown in Tables 12, 13 and 14. To compare the results, the regions chosen were the same as those described for the previous samples: a larger area with grains and plates, regions with grains indicated by an arrow and plates. The regions are represented by Figures 44, 45 and 46 respectively, and the data obtained is contained in Tables 12, 13 and 14, respectively, corresponding to the compounds Bi2.18Sr2.00Ca0.95Cu1.87Ox (total), Bi2.29Sr2.00Ca0.95Cu1.76Ox (grains) and Bi2.10Sr1.93Ca0.88Cu2.08Ox (plates). Once again, we can verify what was observed in the SEM with the BSE detector and check that the desired stoichiometry was obtained quantitatively.

Figura 44. Image of a large area containing grains and plates from EDS analysis for sample Ref - 845°C/72h.

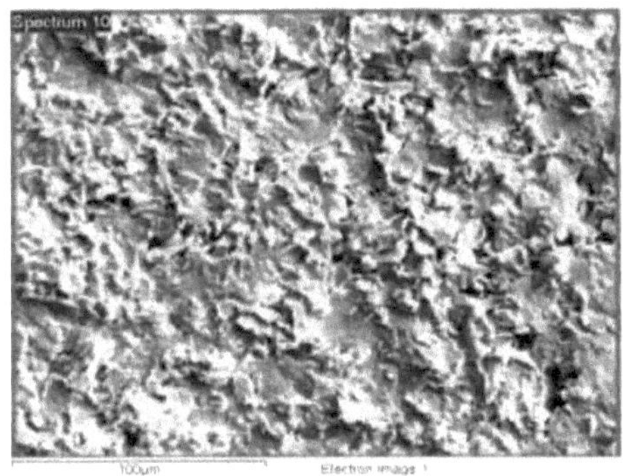

Source: Author's own elaboration.

Table 12. EDS results for sample Ref - 845°C/72h in the area containing grains and plates

Element	Atomic %	Nominal	EDS
Bi	31,12	2	2,18
Mr	28,58	2	2,00
Ca	13,58	1	0,95
Cu	26,76	2	1,87

Source: Author's research data.

Figura 45. Image of the grain region indicated by the arrow, which was analyzed by EDS for sample Ref - 845°C/72h.

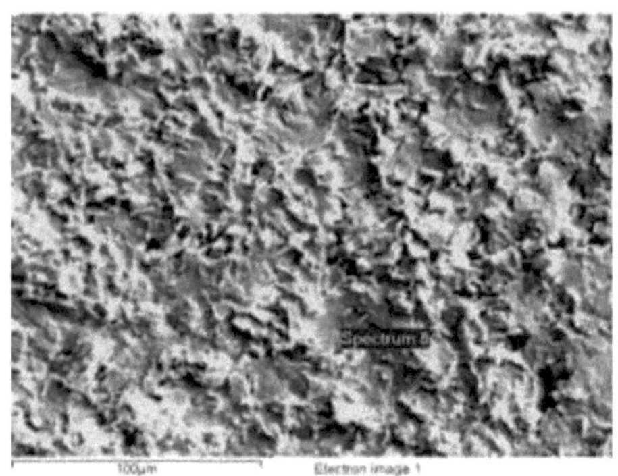

Source: Author's own elaboration.

Table 13. EDS results for sample Ref - 845°C/72h in the grain region.

Element	Atomic %	Nominal	EDS
Bi	32,73	2	2,29
Mr	28,61	2	2,00
Ca	13,53	1	0,95
Cu	25,13	2	1,76

Source: Author's research data.

Figura 46. Image of the plate region analyzed by EDS for sample Ref - 845°C/72h.

Source: Author's own elaboration.

Table 14. EDS results for sample Ref - 845°C/72h in the plaque region.

Element	Atomic %	Nominal	EDS
Bi	30,03	2	2,10
Mr	27,55	2	1,93
Ca	12,54	1	0,88
Cu	29,68	2	2,08

Source: Author's research data.

Again, a comparison was made between the results of the EDS analyses for CaMAH samples treated at 845°C/72h. These data are shown in Tables 15, 16 and 17, and the regions of a larger area containing grains and plaques, grains and plaques, are illustrated in Figures 47, 48 and 49 respectively. The data contained in the Tables for a larger area is $Bi2_{,0}$ $5Sr2_{,19}$ Ca $,_{10}$ 5Cu , O_{170x} , grain analysis is Sr $,_{160}$ Ca $,_{085}$ Cu , O_{485x} and plaques is Bi $,_{219}$ Sr $,_{196}$ Ca $,_{087}$ Cu , O_{196x} . For the analyzed regions of larger area and plaques, it can be seen that the stoichiometries are close to the desired ones and in grain regions, as already predicted in the SEM with BSE detector, the stoichiometry is different. This result is also in line with the literature (BAGHURST; CHIPPINDALE; MINGOS, 1988), which tells us that for microwaves the heating takes place inside the copper oxides and is transferred from there to the surrounding areas.

Figura 47. Image of a larger area with grains and plates from the EDS analysis of the CaMAH sample - 845°C/72h.

Source: Author's own elaboration.

Table 15. EDS results for the CaMAH - 845°C/72h sample in the larger area containing grains and plates.

Element	Atomic %	Nominal	EDS
Bi	29,28	2	2,05
Mr	31,38	2	2,19
Ca	15,51	1	1,05
Cu	24,33	2	1,70

Source: Author's research data.

Figura 48. Image of grain region indicated by arrows from EDS analysis for CaMAH sample - 845°C/72h.

Source: Author's own elaboration.

Table 16. EDS results for the CaMAH - 845°C/72h sample in the grain region.

Element	Atomic %	Nominal	EDS
Bi	-	2	-
Mr	31,38	2	1,60
Ca	11,72	1	0,82
Cu	65,39	2	4,58

Source: Author's research data.

Figura 49. Image of the plaque region, indicated by the arrow, taken by EDS analysis for the CaMAH - 845°C/72h sample.

Source: Author's own elaboration.

Table 17. EDS results for the CaMAH - 845°C/72h sample in the plaque region.

Element	Atomic %	Nominal	EDS
Bi	31,26	2	2,19
Mr	28,05	2	1,96
Ca	12,67	1	0,87
Cu	28,03	2	1,96

Source: Author's research data.

The results of the EDS analyses for NiMAH samples heat-treated at 845°C/72h are shown in Tables 18, 19 and 20. Again, the large-area regions containing grains and plates, grains and plates, are shown in Figures 50, 51 and 52 respectively, the data for which are presented in the Tables. For a larger area, the phase $Bi_{2.25}Sr_{1.96}Ca_{0.62}Cu_{1.96}O_x$ is obtained, for the grain region, $Bi_{2.13}Sr_{2.13}Ca_{0.47}Cu_{1.63}O_x$, and for the plate region, the compound $Bi_{2.29}Sr_{2.37}Ca_{0.73}Cu_{1.60}O_x$. It can be seen that the stoichiometry improved with the heat treatment.

Figura 50. Image of an area containing plaques and grains for EDS analysis of the NiMAH - 845°C/72h sample.

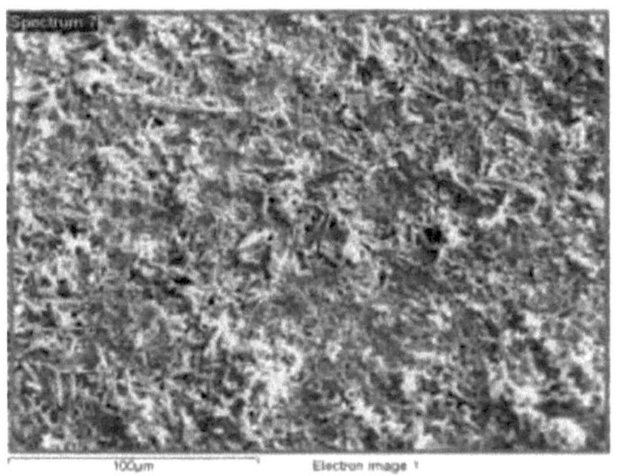

Source: Author's own elaboration.

Table 18. EDS results for the NiMAH - 845°C/72h sample in the area containing plates and grains.

Element	**Atomic %**	**Nominal**	**EDS**
Bi	32,18	2	2,25
Mr	28,05	2	1,96
Ca	9,01	1	0,62
Cu	28,03	2	1,96

Source: Author's research data.

Figura 51. Image of the grain region for EDS analysis of the NiMAH sample - 845°C/72h.

Source: Author's own elaboration.

Table 19. EDS results for the NiMAH - 845°C/72h sample in the grain region.

Element	Atomic %	Nominal	EDS
Bi	30,50	2	2,13
Mr	39,49	2	2,76
Ca	6,69	1	0,47
Cu	23,32	2	1,63

Source: Author's research data.

Figura 52. EDS image of the plate region indicated by an arrow for the NiMAH - 845°C/72h sample.

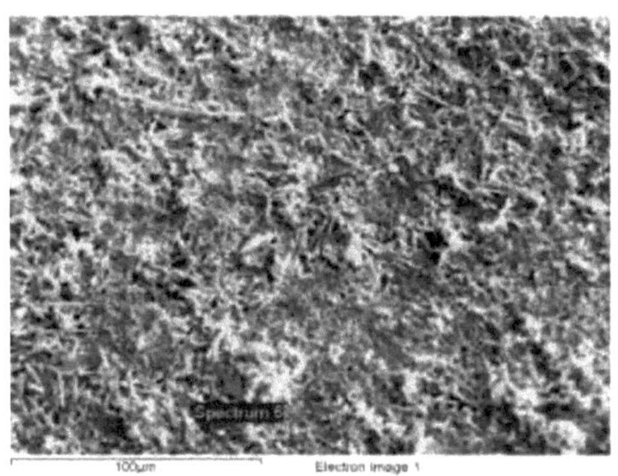

Source: Author's own elaboration.

Table 20. EDS results for the NiMAH - 845°C/72h sample in the plate area region.

Element	Atomic %	Nominal	EDS
Bi	32,74	2	2,29
Mr	33,89	2	2,37
Ca	10,47	1	0,73
Cu	22,90	2	1,60

Source: Author's research data.

To facilitate visualization, a general comparison with the samples, relating region and temperature.

The results in Tables 21, 22 and 23 show the comparison of the samples with the heat treatments for the wide, grain and plate regions respectively.

Table 21. EDS results for sample Ref, CaMAH and NiMAH - comparing the 844°C/24h and 845°C/72h treatments in the broad region.

Sample	845°C/24h	845°C/72h
Ref	$Bi_{2.16}Sr_{1.93}Ca_{0.93}Cu_{2.03}O_{x}$	$Bi_{2.18}Sr_{2.00}Ca_{0.95}Cu_{1.83}O_{x}$
CaMAH	$Bi_{1.93}Sr_{2.05}Ca_{0.94}Cu_{2.08}O_{x}$	$Bi_{2.05}Sr_{2.19}Ca_{1.05}Cu_{1.70}O_{x}$
NiMAH	$Bi_{2.18}Sr_{2.32}Ca_{0.66}Cu_{1.83}O_{x}$	$Bi_{2.25}Sr_{1.96}Ca_{0.62}Cu_{1.96}O_{x}$

Source: Author's research data.

Table 22. EDS results for sample Ref, CaMAH and NiMAH - comparing the 844°C/24h and 845°C/72h treatments in the grain region.

Sample	845°C/24h	845°C/72h
Ref	$Bi_{2.19}Sr_{1.93}Ca_{0.95}Cu_{1.92}O_x$	$Bi_{2.29}Sr_{2.00}Ca_{0.95}Cu_{1.73}O_x$
CaMAH	$Bi_{2.99}Sr_{2.41}Ca_{1.10}Cu_{0.48}O_x$	$Sr_{1.60}Ca_{0.85}Cu_{4.25}O_x$
NiMAH	$Bi_{0.88}Sr_{3.41}Ca_{1.04}Cu_{2.00}O_x$	$Bi_{2.13}Sr_{2.13}Ca_{0.47}Cu_{1.63}O_x$

Source: Author's research data.

Table 23. EDS results for sample Ref, CaMAH and NiMAH - comparing the 844°C/24h and 845°C/72h treatments in the plate region.

Sample	845°C/24h	845°C/72h
Ref	Bi2,09Sr1,88Ca0,93Cu2,09Oχ	Bi2,10Sr1,93Ca0,88Cu2,00Oχ
CaMAH	Bi2,10Sr1,91Ca0,90Cu2,08θχ	Bi2,19Sr1,96Ca0,87Cu1,96θχ
NiMAH	Bi2,07Sr2,84Ca0,44Cu1,61Oχ	Bi2,29Sr2,37Ca0,73Cu1,60Oχ

Source: Author's research data.

EDS analyses are semi-quantitative. Therefore, it cannot only be used to determine stoichiometry; more quantization techniques would be required.

4.4 ELECTRICAL MEASUREMENTS (RXT)

Electrical measurements were carried out using the four-point DC method. Figure 53 shows the resistance curves as a function of temperature (RxT) and also the resistance derivative as a function of temperature (dR/dTxT) for the Ref sample. In general, the T_c's of the samples were obtained using the "*onset*" method (POOLE JR. et al., 2007). For the Ref - 845°C/24h sample, there are two transitions, one at (86 ± 1)K and the other at (81 ± 1)K. For the Ref - 845°C/72h sample, there is a T_c at (85 ± 1)K and a second transition at (79 ± 1)K, i.e. in all samples there is a predominance of the 2212 phase, which confirms the data obtained by XRD and EDS.

Figura 53. (a) Graph of electrical resistance as a function of temperature and (b) the dR/dTxT derivative for the Ref - 845°C/24h and Ref - 845°C/72h samples.

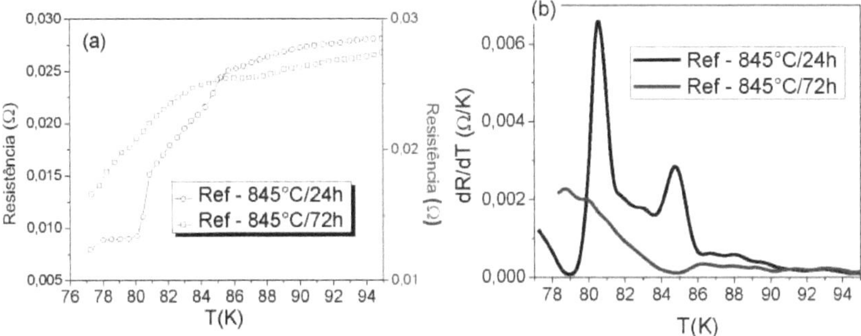

Source: Author's own elaboration.

Figure 54 shows the RxT and dR/dTxT curves for the CaMAH sample. It can be seen that the Tc of the sample treated at 845°C/24h is (84 ± 1)K and the temperature of the second transition is (80 ± 1)K. Figure 57(b) shows noisier curves, but it is still possible to observe the transition. For CaMAH-845°C/72h, there is an increase in Tc to a value of (89 ± 1)K and a second transition at (85 ± 1)K.

Figura 54. (a) Graph of electrical resistance as a function of temperature and (b) the dR/dTxT derivative for the CaMAH - 845°C/24h and CaMAH - 845°C/72h samples.

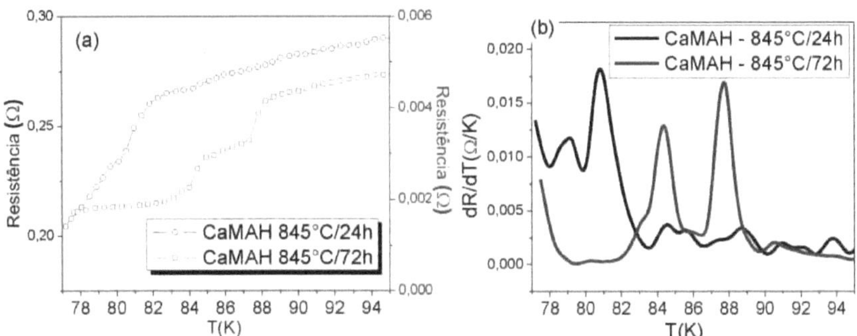

Source: Author's own elaboration.

For the NiMAH sample, the RxT and dR/dTxT curves are shown in Figure 55. NiMAH-845°C/24h showed a T_c at (92 ± 1)K, and other transitions at (90 ± 1)K, (85 ± 1)K and (80 ± 1)K. The NiMAH-845°C/72h sample has a T_c of (89 ± 1)K, with transitions at (85 ± 1)K and (77 ± 1)K.

Figure 55. (a) Graph of electrical resistance as a function of temperature and (b) the dR/dTxT derivative for NiMAH - 845°C/24h and NiMAH- 845°C/72h samples.

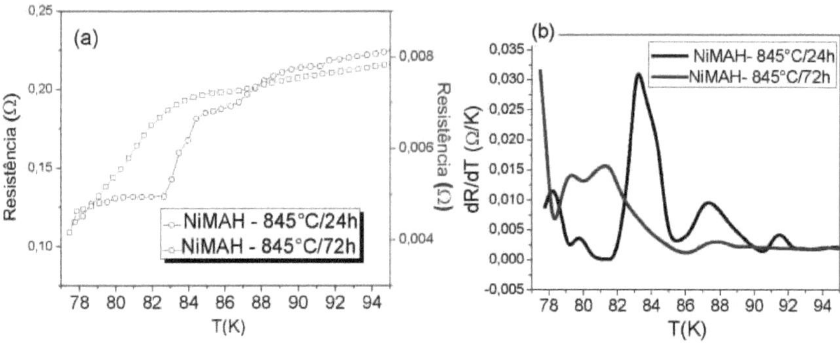

Source: Author's own elaboration.

As can be seen in Figures 53, 54 and 55, all the samples show more than one transition. This behavior may be associated with some factors linked to the inhomogeneous distribution of the grains in the material, such as the appearance and/or disappearance of intergranular currents in some grains; other superconducting phases present in the sample that have different T_c's from the main matrix or, even, the same phase, but with inappropriate oxygen stoichiometry, which is responsible for a worsening of the material's superconducting properties (POOLE JR. et al., 2007).

The notable difference between the responses of the CaMAH and NiMAH samples may indicate that the type of reagent, i.e. nitrate and/or oxide, can influence the final sample obtained when using the microwave-assisted hydrothermalization process.

4.5 MAGNETIC measurements

4.5.1 Magnetization X Temperature (MxT)

The MxT curves in Figure 56 also confirm that the samples mostly have the 2212 phase, as the T_c 's are around 79 K. Another feature that can be seen in Figure 56 (a) and (b) is that the FC curves of the two samples are similar. However, the ZFC curve for the Ref sample shows a greater diamagnetic response, in module, than the CaMAH sample. This is an indication that the grains in the CaMAH sample have less effective weak-links (MOTTA, 2009; POOLE JR. et al., 2007), i.e. it is more difficult for intergranular currents to tunnel between the grains in the sample. For the NiMAH sample, we can see that as the magnetic field increases from 10 Oe to 100 Oe, its response is less diamagnetic and there is a greater loss of superconductivity when compared to the Ref and CaMAH samples. No magnetic measurements were made for samples heat-treated at 845°C/72h, as there was not enough time to do so.

Figura 56. MxT measurements for the samples (a) Ref, (b) CaMAH and (c) NiMAH, showing the ZFC and FC curves for 10 and 100 Oe, and the critical temperatures for each sample.

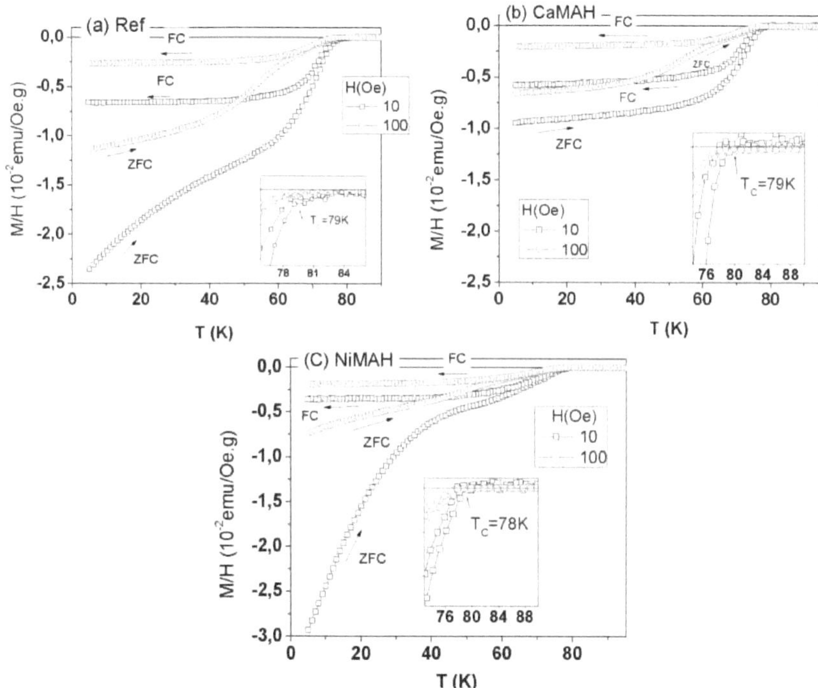

Source: Author's own elaboration.

4.5.2 AC susceptibility measurements

AC susceptibility measurements aim to obtain more information about the magnetic behavior of samples and are a powerful tool for analyzing the response of inter- and intragranular currents. The latter are those that surround each grain, shielding them from the external field. Information can also be obtained on the presence or absence of other superconducting phases (POOLE JR. et al., 2007). In these measurements, two fields with amplitudes an order of magnitude apart were used.

Figure 57 shows the differences between the Ref, CaMAH and NiMAH samples. The CaMAH sample shows a second transition at 53 K, which was only present in the measurement carried out with an excitation field of h=0.01 Oe. This means that, for this field strength, above 53K the intergranular currents cease to exist. However, for both fields, the sample transitions at T_c = 78K. The imaginary susceptibility, χ'' (positive curve), indicates the energy dissipated in the material by the flow in and out of its interior due to the application of an oscillating field. Thus, the wider the peak of this component, the greater the distribution of intergranular currents and, therefore, the more sensitive the sample becomes to the application of an AC field.

For the Ref sample, Tc = 78 K and a second, smoother transition occurs at 63 K for h=1.0 Oe. The Tc indicates a predominance of the 2212 phase, and the second transition may also be related to intergranular currents. From the χ" response, which shows a narrower peak, it can be seen that the superconducting properties of this sample are better compared to the other samples studied, or at least there is greater homogeneity of the weak-links. The NiMAH sample shows a second transition around 63 K for h=0.01 Oe and around 59 K for h=1.0 Oe. The temperature dependence of this second transition is a strong indication that it refers to intergranular currents, because if it referred to another superconducting phase, this transition would refer to the superconducting grain and therefore, like Tc, it would occur at the same temperature, regardless of the h field. As can be seen, then, the Tc of NiMAH is 77 K and is independent of the excitation field. The χ" response suggests that this sample has better intergranular properties than the CaMAH sample and very close to those of the Ref sample.

Figure 57. Measurement of χ_{ac} vs. T. for the samples (a) Ref, (b) CaMAH and (c) NiMAH using a frequency of 1 kHz and amplitudes of 0.01 and 1.00 Oe.

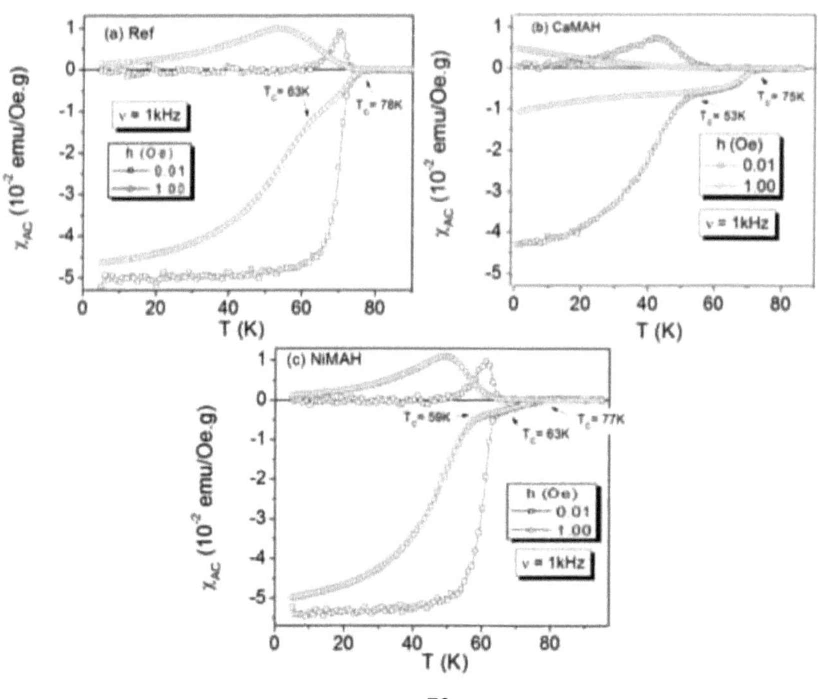

Source: Author's own elaboration.

5 CONCLUSIONS

In this work, three samples were synthesized using different methods and reagents, i.e. a reference sample, Ref, was produced using the Pechini method in which the oxides of the components required to obtain Bi-Sr-Ca-Cu-O were used as reagents. A second sample, called CaMAH, was also synthesized using oxides, but the Pechini method was combined with the microwave-assisted hydrothermalization method, MAH. The third sample, called NiMAH, was synthesized following the same procedures as CaMAH, but the nitrates of the BSCCO components were used. From the XRD, EDS, electrical and magnetic characterizations, it can be seen that the desired phase, 2212, was formed predominantly in all the samples. It is important to note that, in the case of this work, the microwave-assisted hydrothermalization method requires less time to synthesize than the Pechini method that is usually used to prepare ceramic samples such as BSCCO. This usual procedure takes around 20 days, while using the microwave-assisted hydrothermalization method, the solution preparation time was just 2 hours.

Although the XRD analyses showed the predominance of the 2212 phase in all the samples, both MAH and NiMAH had segregated phases.

The SEM images showed the presence of spherical-shaped grains along with plate-shaped grains in the samples produced by the MAH method. As expected, when using the backscattered electron detector, the plates, which are characteristic of BSCCO, show a homogeneity of color that is different from the color of the spherical grains. This is a strong indication that these formations have a different composition to the matrix and may not even be made of superconducting material. It is possible that this result is associated both with the interaction of the microwaves with the material, particularly with the copper oxide, and with the rapid heat transfer that they produce in the material.

The EDS analyses confirmed some of the results of the XRD and SEM, i.e. the samples have stoichiometries of the 2212 phase, and it was possible to confirm that, in the plate region, the stoichiometry was also of the superconducting 2212 phase. The samples produced using the MAH method, on the other hand, had spherical grains with a different stoichiometry to the 2212 phase, i.e. these grains were from other phases and, consequently, influenced the superconducting properties of these materials.

The electrical measurements of the samples also confirmed the 2212 phase; however, they show various transitions which are possibly due to the heterogeneity of the material and the formation of other superconducting phases. The NiMAH samples, in their different heat treatments, showed the highest T_c's.

Another round of heat treatment, over a longer period of time, showed that this had an impact on the samples. In the case of the Ref sample, the segregated phases decreased, but the electrical response worsened. For the CaMAH sample, the heat treatment improved both the electrical and morphological properties of the sample. For the NiMAH sample, there was a morphological improvement observed by XRD, EDS and SEM, but the critical temperature decreased. The decrease in Tc may be associated with the loss of oxygen during the new heat treatment, since it was not carried out with a flow of this gas.

The magnetic characterization results showed that the Ref sample has a more intense magnetization modulus than the CaMAH and NiMAH samples and also a narrower distribution of critical currents. The latter is characterized by a narrow χ'' peak. The causes of these responses may be associated with the formation of the grains and weak-links in the material, i.e. fewer segregated phases were produced in the Ref synthesis process, which allows for better homogeneity and, consequently, connectivity between the grains. It can therefore be concluded that the sample produced by the Pechini method showed better results than those produced with MAH. However, when comparing the samples synthesized using the MAH method, the specimen produced from nitrate precursors showed better results. This result can be attributed to the fact that the nitrates are easier to dissolve and, consequently, the samples are purer.

As is well known, the microwave-assisted hydrothermal method has extremely fast crystallization kinetics (ZHU; HANG, 2007), which can influence the formation of the desired phase. However, copper oxide is a good microwave absorber and this can also influence the formation of the final phase. Therefore, further studies on the influence of the MAH method on the manufacture of ceramic superconductors are necessary.

6 FUTURE PROSPECTS

To better study the residence time and heating rate of the microwave-assisted hydrothermal method.

Obtain phase 2223 with the addition of lead in the stoichiometry, because as the hydrothermal method is a closed system, it would make it difficult for the lead element to volatize and thus be able to obtain the phase.

Using the microwave-assisted hydrothermal method with another type of associated synthesis, unlike the Pechini method.

Increase heat treatment time at all stages of the process.

Study the behavior of the critical density, Jc, based on magnetic hysteresis (M-H).

To study the phases and network parameters of the material through structure refinement by the Rietveld method.

REFERENCES

AL-HARAHSHEH, M.; KINGMAN, S. W. Microwave-assisted leaching: a review. **Hydrometallurgy**, Amsterdam, v. 73, n. 3, p. 189-203, jun. 2004.

BAGHURST, D. R.; CHIPPINDALE, A. M.; MINGOS, D. M. P. Microwave syntheses for superconducting ceramics. **Nature**, Oxford, v. 332, p. 311, 1988.

BALSEIRO, C.; CRUZ, F. Superconductivity. **Ciência Hoje**, Sao Paulo, v. 9, n. 49, p. 27-35, dec. 1988.

BEDNORZ, J. G.; MÜLLER, K. A. Possible High-TC superconductivity in the Ba-La-Cu-O system. **Zeitschrift Für Physik B-Condensed Matter**, Zwitzerland, v. 64, n. 2, p. 189-193, 1986.

BRINKER, J.; SCHERER, G. W. **Sol-Gel science**: the physics and chemistry of Sol-Gel processing. Boston: Academic Press, 1989.

BUCKEL, W.; KLEINER, R. **Superconductivity:** fundamentals and applications. 6. ed. Weinheim: VCH, 2004.

BYRAPA, K.; ADSCHIRI, T. Hydrothermal technology for nanotechnology. **Progress in Crystal Growth and Characterization of Materials**, Kidlington, v. 53, p. 117-166, 2007.

doi:10.1016/j.pcrysgrow.2007.04.001

BYRAPA, K.; YOSHIMURA, M. **Handbook of hydrothermal technology.** New Jersey: Noyes, 2001.

CALLISTER JUNIOR, W. D. **Materials science and engineering:** an introduction. 7. ed. Rio de Janeiro: LTC, 2008.

CARDOSO, A. V. **Ciência dos materiais multimidia**. Belo Horizonte: CETEC/LMDM, 2012. Available at: <http://www.cienciadosmateriais.org>. Accessed on: August 20, 2012.

CARDOSO, C. A. **The AC susceptibility technique applied to the study of type II superconductors.** 2001. 143 f. Thesis (Doctorate in Physics) - Instituto de Fisica "Gleb Wataghin", Universidade Estadual Campinas, Campinas, 2001.

CHU, C. T.; DUNN, B. Preparation of high-tc superconducting oxides by the amorphous citrate process. **Journal of the American Ceramic Society,** Ann Arbor, v. 70, n. 12, p. C-375- C-377, 1987.

CYROT, M. Ginzburg-Landau theory for superconductors**. Reports Progress Physics**, Bristol, v. 36, p. 103-158, 1973.

DEDAVID, B. A.; GOMES, C. I.; MACHADO, G. **Scanning electron microscopy**: applications and sample preparation: polymeric, metallic and semiconductor materials. Porto Alegre: EDIPUCRS, 2007.

DIAS, L. G. **High critical temperature superconductors**: 25 years. Sao Paulo: Blog Fisica, futebol e falàcias: um pouco de cada, July 25, 2011. Available at: <http://fisicafutebolfalacias.blogspot.com.br/2011/07/supercondutores-de-alta-temperatura.html>. Accessed on: April 30, 2012.

GUARANY, C. A. **Ferroelectric structures**. [S.l.]: Ferroelétricos.com, Dec. 2008. Available at: <http://ferroeletricos.com/perovskita.html>. Accessed on: May 1, 2012.

HOFFMAN, J. **Cuprate supeconductors**. Cambridge: Hoffman Lab, 2010. Available at: <http://hoffman.physics.harvard.edu/materials/Cuprates.php>. Accessed on: May 1, 2012.

HOFFMAN, J. **Scanning tunneling microscopy**. Cambridge: Hoffman Lab, 2012. Available at: <http://hoffman.physics.harvard.edu/research/STMresearch.php>. Accessed on: April 30, 2012.

IYER, A. N.; LU, W.; MIRONOVA, M.; VIPULANANDAN, C.; BALACHANDRAN, U.; SALAMA, K. Current transport and microstructural development in BSCCO tapes and joints fabricated by groove rolling. **Superconductor Science and Technology**, Bristol, v. 13, n. 2, p. 187-194, 2000.

KAKIHANA, M. Sol-gel preparation of high temperature superconducting oxides. **Journal of Sol-Gel Science and Technology,** New York, v. 6, n. 1, p. 7-55, 1996.

KAMIHARA, Y.; WATANABE, T.; HIRANO, M.; HOSONO, H. Iron-based layer superconductor La[$O_{1-X}F_X$]FeAs (x = 0.05-0.12) with Tc = 26 K. **Journal of the American Ceramic Society,** Ann Arbor, v. 130, n. 11, p. 3296-3297, 2008.

KATSUI, A.; OHTSUKA, H. Solution growth of Bi-Sr-Ca-Cu-O compounds using alkali chlorides. **Journal of Crystal Growth**, Amsterdam, v. 91, n. 3, p. 261-263, 1988.

KEYSON, D.; LONGO, E.; VASCONCELOS, J. S.; VARELA, J. A.; ÉBER, S.; DERMADEROSIAN, A. Synthesis and processing of ceramics in a domestic microwave oven. **Ceràmica,** Sao Paulo, v. 52, n. 321, p. 50-56, 2006.

KITTEL, C. **Introduction to solid state physics.** 8. ed. Rio de Janeiro: LTC, 2005.

KOMARNENI, S.; ROY, R.; LI, Q. H. Microwave-hydrothermal synthesis of ceramic powders. **Materials Research Bulletin**, Kidlington, v. 27, n. 12, p. 1393-1405, 1992.

MAEDA, H.; TANAKA, Y.; FUKUTUMI, M.; ASANO, T. A new high Tc Superconductor without a rare earth element. **Journal of Applied Physics,** College Park, v. 27, n 2, p. L209-L210, 1988.

MAJEWSKI, P. Materials aspects of the high-temperature superconductors in the system Bi2O3-SrO-CaO-CuO. **Journal of Materials Research**, New York, v. 15, n. 4, p. 854-870, 2000.

MARTINEZ, L. G. **Study of the crystal structure of the superconducting compound Hg_{1-x} Re_x Ba C_{2a2} Cu $O_{38+⅛}$ - Hg, Re-1223**. 2005. 139 f. Thesis (Doctorate in Sciences) - Nuclear and Energy Research Institute, University of Sao Paulo, Sao Paulo, 2005.

MOREIRA, M. L. **Alkaline earth titanates:** order associated with disorder. 2010. 113 f. Thesis (Doctorate in Sciences) - Federal University of Sao Carlos, Sao Carlos, 2010.

MOTTA, M. **Structural inhomogeneities in nanoscopic samples of high critical temperature superconductors.** 2009. 123 f. Dissertation (Master's Degree in Materials Science and Technology) - Faculty of Sciences, Universidade Estadual Paulista, Bauru, 2009.

OSTERMANN, F.; PUREUR, P. **Superconductivity**. Sao Paulo: Livraria da Fisica: Sociedade Brasileira de Fisica, 2005.

PECHINI, M. P. **Method of preparing alkaline earth titanates and niobates and coating method using the same to form a capacitor**. US Patent n. 3,330,697, Nov. 7, 1967.

PENG, Z. S.; HUA, Z. Q.; LI, Y. N.; DI, J.; MA, J.; CHU, Y. M.; ZHEN, W. N.; YANG, Y. L.; WANG, H. J.; ZHAO, Z. X. Synthesis and properties of the bi-based superconducting powder prepared by the pechini process. **Journal of Superconductivity**, New York, v. 11, n. 6, p. 749-754, 1998.

POOLE JR., C. P. et al. **Superconductivity**. 2. ed. Amsterdam: Elsevier, 2007.

RAO, C. N. R.; NAGARAJAN, R.; VIJAYARAGHAVEN, R. Synthesis of cuprate superconductors.

Superconductor Science and Technology, Bristol, v. 6, n. 1, 22 p., 1994.

RODRIGUES, V. D. **Effects of doping on the electrical properties of the BSCCO superconducting system with rare earth element.** 2011. 74 f. Dissertation (Master's Degree in Materials Science) - Faculty of Engineering, Universidade Estadual Paulista, Ilha Solteira, 2011.

SALAMATI, H.; KAMEL, P.; AKHAVAN, M. Fabrication of BSCCO thin films using sputtering technique. **Physica Status Solidi (C)**, Weinheim, v. 1, n. 7, p. 1895-1898, 2004.

TARASCON, J. M.; LE PAGE, Y.; BARBOUX, P.; BAGLEY, B. G.; GREENE, L. H.; MCKINNON, W.

R.; HULL, G. W.; GIROUD, M.; HWANG, D. M. Crystal substructure and physical properties of the superconducting phase Bi_4(Sr, Ca)6Cu4 O_{16+x}. **Physical Review B (Condensed Matter and Materials Physics)**, College Park, v. 37, n. 16, p. 9382-9389, 1988.

VOLANTI, D. P.; CAVALCANTE, L. S.; KEYSON, D.; LIMA, R. C.; MOURA, A. P.; MOREIRAR, M. L.; MACARIO, L. R.; GODINHO, M. Nanostructured materials obtained by microwave-assisted hydrothermal synthesis. **Metalurgia & Materiais**, Sao Paulo, v. 63, p. 351-357, jul. 2007.

WONG-NG, W. Phase diagrams. In: HANDBOOK of superconductivity. San Diego: Academy Press, 2000. p. 625-685.

ZHANG, Y.; YANG, H.; LI, M.; SUN, B.; QI, Y. Improvement of multiple oxide properties: effect of gel processes on the quality of Bi2Sr2CaCu2O8+δ superconducting powders.

CrystEngComm, Cambridge, v. 12, n. 11, p. 3046-3051, 2010.

ZHU, X. H.; HANG, Q. M. Microscopically and physical characterization of microwave and microwave-hydrothermal synthesis products. **Micron**, Kidlington, v. 44, p. 21-24, Jan. 2013. Available at: <http://www.sciencedirect.com/science/article/pii/S0968432812001795>. Accessed on: Feb. 15, 2013.

Printed by Books on Demand GmbH, Norderstedt / Germany